Electronics Simplified

Electronics Simplified
Previously published as Electronics Made Simple

Third edition

Ian Sinclair

ELSEVIER

AMSTERDAM • BOSTON • HEIDELBERG • LONDON • NEW YORK • OXFORD
PARIS • SAN DIEGO • SAN FRANCISCO • SINGAPORE • SYDNEY • TOKYO

Newnes is an imprint of Elsevier

Newnes

Newnes is an imprint of Elsevier
The Boulevard, Langford Lane, Kidlington, Oxford, OX5 1GB, UK
30 Corporate Drive, Suite 400, Burlington, MA 01803, USA

First edition 1997
Second edition 2002
Third edition 2011

Notices
Knowledge and best practice in this field are constantly changing. As new research and experience broaden our understanding, changes in research methods, professional practices, or medical treatment may become necessary.

Practitioners and researchers must always rely on their own experience and knowledge in evaluating and using any information, methods, compounds, or experiments described herein. In using such information or methods they should be mindful of their own safety and the safety of others, including parties for whom they have a professional responsibility.

To the fullest extent of the law, neither the Publisher nor the authors, contributors, or editors, assume any liability for any injury and/or damage to persons or property as a matter of products liability, negligence or otherwise, or from any use or operation of any methods, products, instructions, or ideas contained in the material herein.

British Library Cataloguing in Publication Data
A catalogue record for this book is available from the British Library

Library of Congress Control Number: 2011920147

ISBN: 978-0-08-097063-9

For information on all Newnes publications
visit our website at www.elsevierdirect.com

Printed and bound in the United Kingdom

11 12 13 14 10 9 8 7 6 5 4 3 2 1

Working together to grow
libraries in developing countries

www.elsevier.com | www.bookaid.org | www.sabre.org

ELSEVIER   BOOK AID
            International   Sabre Foundation

Contents

Preface

The success of the first two editions of this book has shown the continuing interest in and need for a book that deals with the aims and methods of electronics without going into too much detail about practical circuitry, and with the minimum of simple mathematics. This third edition has been prepared with much more emphasis on digital electronics, now that virtually every device available to the consumer is partly or entirely digital. The information on microcomputers has been greatly expanded, with a separate chapter on software options.

There is an old precept that it is easier to teach digital electronics to anyone who knows about linear electronics than the other way round. This does not apply as much as it used to, but the analog (analogue in the UK) fundamentals are still important and they have not been dropped in this edition. For advanced information or for details of new devices the reader is urged to make full use of Internet search engines. For readers who are interested in electronics construction and circuitry, please take a look at *Practical Electronics Handbook*, currently in its sixth edition in the UK.

I hope that this third edition will be received as enthusiastically as were the first and second editions, and that it is equally successful in stimulating interest in this fascinating branch of engineering.

Ian Sinclair

CHAPTER 1

Electricity, Waves, and Pulses

Fundamental Electricity

Definition

Electrical engineering is the study of generation of electrical power from other forms of power (mainly heat, but now including mechanical systems such as wave and wind), its transmission from one place to another, and its use, both industrial and domestic. Electronics is the branch of electrical engineering concerned with the control of individual particles called electrons whose movements we call electric current.

Electrical engineering is concerned with making use of electricity as a way of transmitting and using power. There are natural sources of electrical power, such as lightning and some living creatures, such as the electric eel, but the main concern of electrical engineering is directed to man-made systems that generate, distribute, and use electricity. Our main uses of electricity depend on converting electrical power to other forms, mainly into mechanical effort, heat and light. These other forms are also our main sources of electrical power.

The first electrical effects to be discovered were those of **static electricity**, observed when amber was rubbed with silk. The ancient Greeks first discovered these effects, and because amber is called **electron** in Greek, we have coined the words such as **electricity** and **electronics** that are so familiar today. Considerably later, in the eighteenth century, experimenters discovered other effects that also seemed to be electrical, but these were entirely different. Typically, these effects use a chemical action to generate what we now call electric current in a closed path that we call a **circuit**. As an aid to imagining what was happening, electric current can be compared to the flow of water in pipes.

Definition

Electric current is the flow of electricity through a metal such as a cable. Electric voltage is the electrical form of pressure that forces the current to flow.

Think for a moment about a water circuit, such as is used in central heating. The path for the water is closed (Figure 1.1) and the water is moved by using a pump. No water is lost from the

Electronics Simplified. DOI: 10.1016/B978-0-08-097063-9.10001-9

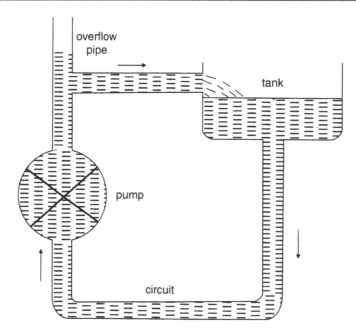

Figure 1.1:
A water circuit, which behaves in many ways like an electrical circuit

circuit and none is added. Turning off a tap at the tank (breaking the circuit) would make the water level rise, because of the pressure of the pump, in the vertical (overflow) piece of pipe. The pump maintains the pressure that makes the water move, and we can use this movement to transmit power because the flowing water can turn a turbine wheel at some other part of the circuit. Hydraulic machines depend on this idea of a liquid in a circuit, though the modern development of hydraulics came later than our use of electricity.

It is not surprising, then, that early experimenters thought that electricity was some kind of invisible liquid. Static electricity effects were explained as being caused by the pressure of this liquid, and current electricity by the flow of the liquid. Electrical engineering is mainly concerned with flow, but the later science of electronics has been based as much on static electricity as on current flow, because we now know much more about what is flowing and why it can flow so easily in metal wires. The fundamental quantity of electricity is electrical charge, and that is what moves in a circuit.

Summary

Electrical effects were once thought to be caused by an invisible liquid that could flow through metals and accumulate on non-metallic materials. Electrical engineering is concerned with the effects of electrical flow, and most of our early applications of electricity have used the effects of flowing electricity.

Definition

All of the effects we call electrical are due to **electric charge**. **Electrostatic** (static electricity) effects are caused by charge at rest, and **electric current** effects (including magnetism) are caused by charge that is moving.

That definition does not tell you much unless you know something about electric charge. No-one knows precisely what charge **is** (though we are slowly getting there), but we do know a lot about what charge **does**, and what we know about what charge does is knowledge that has been accumulated since the time of the ancient Greeks. We can summarize what charge does (the **properties** of charge):

- When you rub two non-metallic objects together they will both usually become charged.
- These charges are of two opposite types, one called positive, the other called negative.
- Two charges of the same type (two positives or two negatives) repel each other; two opposite charges (one positive, one negative) attract each other.
- The natural state of any substance is not to have any detectable charge, because it contains equal quantities of positive and negative charges.

What we know about the way that charge behaves has led to finding out more about what it is, and we know now that charge is one of the most important effects in the Universe. Like gravity, charge is a way of distorting space, so that it appears to cause force effects at a distance from the cause of the charge. What we call charge is the effect of splitting atoms, separating small particles called **electrons** from the rest of each atom. Each electron is negatively charged, and the amount of charge is the same for each electron. The other main part of an atom, the **nucleus**, carries exactly as much positive charge as the electrons around it carry negative charge (Figure 1.2), if we picture the atom as looking like the sun and its planets. For example,

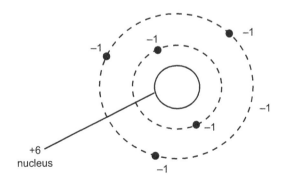

Figure 1.2:
A sun-and-planet view of an atom. Though as a concept this is out of date it helps to illustrate the idea of electrons whose total charge balances the charge of the nucleus and also the idea that the outermost electrons can be detached

if there are six electrons then the nucleus must carry six units of positive charge, exactly balancing the total negative charge on the electrons.

■ Note

Modern physics has long abandoned pictures of atoms as sun-and-planet systems, but this type of picture of the unimaginable is good enough for all purposes concerned with electrical engineering, and for most concerned with electronics.

■

When an electron has become separated from the atom that it belongs to, the attraction between the electron and its atom is, for such tiny particles, enormous, and all the effects that we lump together as electricity, ranging from lightning to batteries, are caused by these force effects of charge. The forces between charges that are at rest are responsible for the effects that used to be called static electricity (or **electrostatics**), and these effects are important because they are used in several types of electronics devices.

■ Note

The forces are so enormous that we can usually separate only one electron from a nucleus, and we can separate all of the electrons only at enormous temperatures, such as we find within the sun or in an exploding hydrogen bomb (which is what the sun actually is).

■

Another option for an electron that has become separated from an atom is to find another atom that has lost its electron (and is therefore positively charged). The movement of electrons from one atom to another causes a large number of measurable effects such as electric current, magnetism, and chemical actions like electroplating. Of these, the most important for electronics purposes are electric current, and one of its effects, magnetism. Materials that allow electrons to move through them are called **conductors**; materials that do not allow electrons to move easily through them are called **insulators**.

■ Note

In some types of crystals, compressing the crystal will separate charges, generating a high voltage. These crystals are termed **piezoelectric**, and a typical application is as an ignition device for a gas fire or cooker. The effect is also reversible, so that a piezoelectric crystal can be used to convert an electrical pulse into a mechanical compression or expansion of the crystal. This effect is used in ultrasonic cleaners.

■

The movement of electrons that we call electric current takes place in a circuit, a closed path for electrons that has been created using conducting material. All circuits for current are closed circuits, meaning that electrons will move from a generator through the circuit and back to the generator again. This is essential because unless electrons moved in a closed path like this many atoms would be left without electrons, and that condition could not exist for long because of the large forces that draw the electrons back to the atoms.

Definition

Electric current is the amount of charge that passes per second any point in a circuit. Electric **voltage** is the amount of work that each charge can do when it moves.

These are formal definitions. We cannot easily count the number of electrons that carry charge along a wire, and we cannot easily measure how much work is done when a charge moves. We can, however, measure these quantities by making use of the effects that they cause. Current along a wire, for example, will cause a force on a magnet, and we can measure that force. The voltage caused by some separated charges can be measured by the amount of current that will flow when the charges are allowed to move. The unit of current is called an **ampere** or **amp**, and the unit of voltage is a **volt**. These terms come from the names of the pioneers Ampere and Volta, and the abbreviations are A and V, respectively.

■ **Note**

We can create less formal definitions for ourselves. Voltage is like a propelling force for current, and current itself can be thought of as like the current of a river. If we continue with this idea, voltage corresponds to the height of the spring where the river starts.

■

All substances contain electrons, which we can think of as being the outer layer of each atom. Some materials are made out of atoms that hold their electrons tightly, and when electrons are moved out of place it is not easy for them to return to their positions. In addition, other electrons cannot move from their own atoms to take up empty places on other atoms. We call these materials **insulators**, and they are used to prevent electric current from flowing. In addition, insulators can be charged and will remain charged for some time.

A good example is the party balloon which is charged by rubbing it against a woolen sweater and which will cling to the wall or the ceiling until its charge is neutralized. Surprisingly high voltages can be generated in this way on insulators, typically several kilovolts (kV), where kilo means one-thousand. For example, 5 kV means five-thousand volts. A very small current can discharge such materials, and we use the units microamp (μA), meaning a millionth of an amp, nanoamp (nA), meaning a thousandth of a millionth of an amp, and picoamp (pA), meaning a millionth of a millionth of an amp. There is an even smaller unit, the femtoamp (fA), one-thousandth of a picoamp.

■ Note

As a comparison, the electrical supply to a house in the UK is at 240 V ± 10%, 50 Hz and currents of 1 A to 13 A are used in domestic equipment (though electric cookers can use up to 30 A). In the USA, the minimum supply voltage is 110V to 115V, and the maximum is 120V to 125V, at 60 Hz, with higher currents (requiring thicker wiring).

■

Now let's look at electricity with lower voltages and higher currents. This is the form of electricity that we are most familiar with and which we use daily.

Steady Voltage

There are several ways of generating a steady voltage, but only two are important for everyday purposes. Batteries are the most familiar method, and the invention of the first battery by Alessandro Volta in 1799 made it possible for the first time to study comparatively large electric currents at voltage levels from 3 V to several hundred volts. A battery is, strictly speaking, a stack of **cells**, with each cell (Figure 1.3) converting chemical action into electrical voltage. In the course of this, a metal is dissolved into a metal-salt, releasing energy in the form of electrical voltage that can make current flow.

■ Note

If the released energy were not converted into electrical form it would be converted to heat, which is what most chemical changes provide. As it is, if you take a large current from a battery you will find that it gets hot.

■

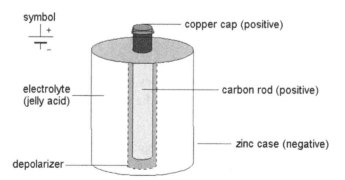

Figure 1.3:
A typical cell, the familiar zinc–carbon type. The voltage output is 1.5 V for a fresh cell, and the current that can be drawn depends on the size of the cell, up to a few amperes. The electrical energy is obtained at the expense of chemical energy. The acid dissolves the zinc case, releasing electrons so that the case is negative. The charge accumulates on the carbon rod. The depolarizer removes hydrogen gas which otherwise acts as an insulator

For centuries the zinc—carbon type of cell, and a few others, were the only types known, but in the later twentieth century several other types were discovered. One of these, the lithium-ion (often abbreviated to **L-ion**), is totally different in construction and delivers more than 3 V, unlike other types that provide a maximum of 1.5 V. Just to confuse matters, there is also a lithium cell (*not* lithium-ion) that provides just 1.5 V but has a much longer useful life than the older types.

Some types of cell, such as Li-ion and nickel—metal hydride (NiMH), are rechargeable, so that you can connect them to a (higher) voltage and convert the metal-salt back to the metal — but you need to use more energy than you got out of the cell. As always, no energy is ever created. If anyone ever tries to sell you a motor that runs on air, magnetism, or moonbeams, always ask where the energy comes from (usually it is from the people who have been cheated along the way). You can always detect a charlatan by the way that he or she uses spoof-science phrases like *tapping into our natural energy field* (and they usually bring crystals into their vaporings as well).

Another way of generating a steady voltage was discovered by Michael Faraday in 1817. He demonstrated the first **dynamo**, which worked by rotating a metal disk between the poles of a magnet, using the energy of whatever was turning the disk (Faraday's hand first of all, and later a steam engine) converted into electrical voltage that could provide current (Figure 1.4). Strictly speaking, a voltage that is the result of a generator or a battery should be called an electromotive force (EMF), but this name is slightly old-fashioned nowadays.

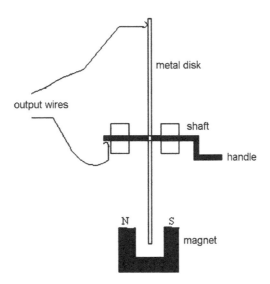

Figure 1.4:
Faraday's first electrical generator. When the disk is spun, a voltage (EMF) appears between the contact points or brushes. The voltage is very small, but the principle can be applied to make a dynamo

■ Note

Later, Faraday found that higher voltages could be generated by substituting a coil of wire for a metal disk, and this is the basis for modern generators, though the original Faraday principle (the **homopolar dynamo**) is still in use for specialized applications.

■

In these pioneering days, steady (or **direct**) voltage was the only type that was thought to be useful. A steady voltage will cause a steady current to flow when there is a path of conducting material, a **circuit**, between the points where the voltage (the EMF) exists. We call these points **terminals**. A cell or a simple dynamo will have two terminals, one that we call **positive** and the other **negative**. When a wire, or any other conductor, is used to connect the terminals, a current will flow, and if the voltage is steady, then the current also will be steady. That does not mean that it will be steady for ever. A cell will be exhausted when its metal is all dissolved, and the voltage will fall to zero. A dynamo will generate a voltage only for as long as the shaft is turned. By convention, we say that the current flows from the positive terminal to the negative terminal (even though we know now that the electrons move in the opposite direction).

A steady voltage will cause a steady current, called direct current (**DC**), to flow, and in 1826 Georg Simeon Ohm found what determined how much current would flow. He called this quantity **resistance**. The three quantities, voltage, current, and resistance, are therefore related, and the unit of resistance is called the **ohm** in his honor. For any part of a circuit, we can measure the voltage across the circuit (between one end and the other) and the current through the circuit and so find the resistance.

Definition

When a current of I amps flows, using a voltage of V tubevolts, then the resistance R ohms is equal to voltage divided by current. In symbols, this is $R = V/I$, and we can write this also as $V = RI$ or $I = V/R$.

■ Note

This relationship is often, wrongly, called **Ohm's law**. The correct definition of Ohm's law states that this quantity called resistance is constant for a metal that is used to conduct current at a steady temperature. What this boils down to is that we can use the relationship in any of its three forms with R constant if the resistance R is of a metal at constant temperature. In other words, if 6 V causes 2 A to flow through a resistance, then 12 V will cause 4 A to flow through the same piece of metal (whose resistance is 3 ohms, written as 3 Ω. This is not true when some non-metal materials are used, or if a metal is allowed to get hot. For example, the current through a semiconductor obeys

$I = V/R$, but the value of R is not constant, it changes for each value of current. Similarly, a torch bulb has a much higher resistance when it is hot than it has when it is cold. We cannot use Ohm's law assuming a constant value of resistance for these examples.

■

We can determine whether or not a conducting material obeys Ohm's law by plotting a graph of current against voltage, using a circuit as in Figure 1.5(a). If this results in a graph that is a straight line (Figure 1.5b), then the material obeys Ohm's law; it is **ohmic**. Most metals behave like this if their temperature is kept constant. If the graph is curved (Figure 1.5c) the material is not ohmic, and this type of behavior is found when metals change temperature considerably or when we use semiconducting materials in the circuit.

Power

Power is the *rate* of doing work, meaning the amount of work done per second. When we use electricity, perhaps for heating, lighting, running a motor, or plating gold on to a base metal, we are making use of power, converting it from the electrical form to other forms. The

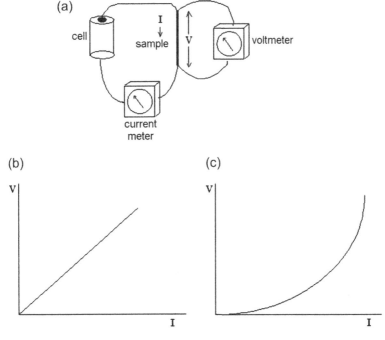

Figure 1.5:
Voltage, current and resistance. (a) The voltage (EMF) of a cell is used to pass current through a sample of conducting material: the ratio of voltage across the sample to current through the sample defines the resistance; (b) ohmic graph; (c) one form of non-ohmic graph

amount of power that is dissipated or converted can be calculated easily for anything that uses steady voltage and current; it is equal to the figure of volts multiplied by the figure of current. When this power is converted into heat we often call it **dissipation** because we cannot contain it; it leaks away. The calculation of power, in symbols, is:

$$\text{Power} = V \times I$$

where V is in volts and I is in amps. The unit of power is the watt, abbreviation W. As an example, if the torch bulb is rated at 3 V 0.5 A then its power is 1.5 W.

Things are not so simple when we are working with quantities that are not steady, but we can always find the amount of power by making measurements of volts and amps and carrying out a multiplication. The difference is that for changing voltages and currents we need to multiply by another factor as well; we will look at that later when we deal with RMS quantities and phase angles.

Alternating Voltage

The voltage (EMF) that is generated by a cell or battery is truly steady, but the voltage from a rotating generator is not. When one side of the revolving coil of wire is approaching a pole of the magnet, the changing voltage is in one direction, but as the wire moves away from the magnetic pole, the changing voltage is in the other direction. Because there are two poles to a magnet, the voltage from a rotating coil rises and falls twice in a revolution, positive for half the time and negative for the other half. This is an **alternating** voltage, alternately in one direction and then the other direction. The graph looks like that of Figure 1.6, and the time it takes to go through one complete cycle of the waveshape is the time that it takes to turn the coil through a complete revolution. The snake shape of the graph gives rise to the name **sinewave**, from the Latin word for 'snake'.

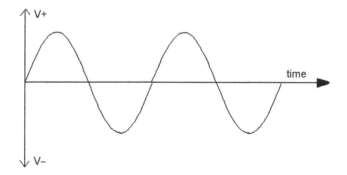

Figure 1.6:
A sinewave that can be produced by rotating a coil between the poles of a magnet. This is the form of AC wave used for electrical supplies and also for radio carrier waves (see later). Radio carrier waves are generated by circuits called oscillators rather than by rotating machines

Faraday's later type of dynamo got around this reversal of current by using a mechanical switch called a **commutator**, which reversed the connections to the coil twice on each revolution, just as the voltage passed through zero (Figure 1.7). This generated an **EMF** which, though not exactly steady, was at least always in the same direction, so that it was possible to label one terminal as plus and the other as minus. If the shaft of the dynamo is spun quickly enough, there is in practice very little difference between the voltage from the dynamo and the same amount of voltage from a battery. For tasks like heating, lighting, electroplating, battery charging, and so on, the supplies are equivalent.

There are other ways of generating electricity from heat and from light, but they suffer the problems of inefficient conversion (not much electrical energy out for a lot of heat energy in) or low density of energy (for example, you have to cover a lot of ground with light cells to generate electricity for a house). A few small generators use nuclear power directly, by collecting the electrons that radioactive materials give out, but large-scale nuclear reactors use the heat of the reaction to generate steam and supply it to turbines. These are therefore steam-powered generators, and the only difference is in how the steam is obtained (and the hysteria that is generated).

The same is true of the places on Earth where steam can be obtained from holes in the ground, providing geothermal power, but this steam-power is less controllable and certainly not available everywhere. Higher efficiency figures can be obtained where heat is not involved, such as in hydroelectric generators. Wind turbines, in contrast, require the variable voltage that they generate to be turned into a constant voltage and frequency (see later) and these conversions reduce the already low efficiency. If we were really serious about spending money

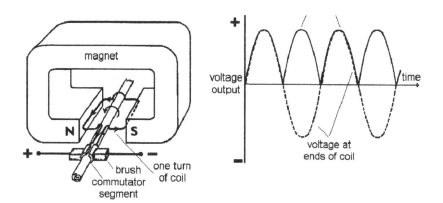

Figure 1.7:
Principle of the dynamo. The coil generates a voltage that is alternating, but by reversing the connections on each half turn, the output voltage is in one direction (a unidirectional voltage). Modern dynamos use slip-rings and semiconductor circuits (see Figure 2.18) rather than the old-fashioned commutator (which wears out because of sparking and mechanical rubbing)

wisely we would try to develop tidal or wave generators that would serve the additional function of protecting our coasts from erosion.

■ Note

Incidentally, we could count electricity generated by solar heating as nuclear, because the Sun, like other stars, is a gigantic nuclear furnace. The difference is that you do not (yet) get demonstrators demanding that the Sun should be shut down or moved. Unlike chemical energy, electricity cannot be stored in any useful quantity so we have to generate as much as we use.

■

Summary

Voltage (EMF) can be generated by converting energy (mechanical, chemical, or heat energy) into electrical form. A cell converts chemical energy into a steady voltage, but only for as long as there is metal to supply the energy. A dynamo uses mechanical energy and will generate a voltage for as long as the shaft can be turned by a steam engine, wind-power, a waterwheel, or whatever source of energy is at hand. The output of a dynamo is not steady, but is in one direction and can be used for the same purposes as truly steady voltage.

Alternating Current and Waves

If we omit the commutator of a dynamo and connect to the ends of the coil rotating at a steady number of revolutions per second, the graph of the voltage, plotted against time, shows a shape that is a sinewave, as was illustrated in Figure 1.6. The connection to the coil can be made using slip-rings (Figure 1.8), so as to avoid twisting the connecting wires. This voltage is an

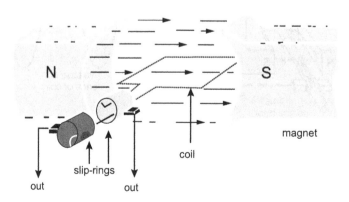

Figure 1.8:
Slip-rings provide a way of connecting wires to each end of a rotating coil

alternating voltage and when we use an alternating voltage in a circuit, the current that flows is **alternating current** (**AC**). A graph of current plotted against time is of the same shape as the graph of voltage, and the peak of current is at exactly the same time as the peak of voltage in each direction if the circuit contains only resistances.

■ Note

In practice, the rotating portion is usually a magnet that uses current in a coil (an **electromagnet**), and the slip-rings carry the steady current that magnetizes this rotor. The output is taken from a coil wound on the non-rotating portion (stator), so that the much larger output current does not have to be passed through a brush and slip-rings. This device is called an **alternator**, and its most familiar form is the generator in your car. The AC output from the alternator is converted into DC to charge the battery.

■

We can use AC for electrical heating, for electric light, for motors, and in fact for most domestic uses of electricity. It cannot be used for electroplating or battery charging, however, and it cannot be used for most types of electronic equipment. In the nineteenth century, when electronics was in its infancy, the advantages of AC greatly outweighed any minor disadvantages, particularly since AC could be converted to DC if DC were essential. What are the advantages of using AC?

- AC is the natural output from a rotating generator, requiring no commutators or other reversing devices.
- Alternating voltage can be converted up or down (using a **transformer**; see later) without using any mechanical actions. For example, if you generate at 5 kV you can convert this up to 100 kV or down to 240 V with negligible losses. The higher the voltage you convert to, the longer the distance you can connect by a cable of reasonable size. This makes the National Grid possible, so that generating stations need not be close to users of electricity.
- Very simple motors can be made that use AC and which will run at a constant speed (used for clocks, gramophone motors, and tape-recorder motors).
- AC can be used to power vibrating motors, such as used for electric shavers.
- AC can be converted to DC for electronic equipment, and can be used to provide several different steady voltage levels from one AC supply.

Virtually every country in the world therefore generates and distributes electricity as AC, and the convention is to use a rate of 50 cycles per second (50 hertz or Hz) in Europe or 60 cycles per second (60 Hz) in the USA. In terms of a simple generator, this corresponds to spinning the shaft of the generator at 3000 revolutions per minute (r.p.m.) for 50 Hz, or 3600 r.p.m. for 60 Hz. The abbreviation Hz is for hertz, the unit of one cycle of alternation per second. This was named after Heinrich Hertz, who discovered radio waves in 1884.

■ Note

Most of Europe uses AC at 240 V, 50 Hz, but US domestic appliances need thicker cables than their European counterparts (for the same amount of power) because a higher current is needed to provide the same power at the lower voltage.

■

Summary

DC is the natural output of a battery, but AC is the natural output of a rotating generator. AC is used worldwide for generating and distributing electricity, mainly because it makes it possible to have a large distance between the generator and the user. Since AC can be converted to DC much more easily than converting DC to AC, there are no problems in using an AC supply for electronics circuits that require a steady voltage supply.

Electronics

Definition

Electronics is a branch of electrical engineering that is concerned with controlling charged particles such as electrons and holes. We will introduce the idea of **holes** now.

Atoms such as those of metals can be so tightly packed together that they can share electrons, and the loss of an electron from a set of these packed atoms does not cause such a large upset in any one atom. In these materials, electric current can flow by shifting electrons from one set of atoms to the next. We call these materials **conductors**. All metals contain closely packed atoms, and so all metals are conductors, some much better than others. A small voltage, perhaps 1.5 V, 6 V, 12 V or so, can push electrons through a piece of metal, and large currents can flow. These currents might be of several amps, or for very low voltages perhaps smaller amounts (see earlier) measured in milliamps (mA), microamps (µA), picoamps (pA), or even femtoamps.

■ Note

Pure liquids, other than liquid metals such as mercury or gallium, are not good conductors, but liquids with dissolved salts will conduct because when a salt dissolves in water (for example), the solid salt is split into charged particles called **ions**. These ions can move, so that the solution will conduct electricity. Note that the word 'salt' means any compound made by combining a metal and a non-metal, of which common salt (sodium chloride) is the most familiar example.

■

As well as being closely packed, the atoms of a metal are usually arranged in a pattern, a **crystal**. These patterns often contain gaps in the electron arrangement, called **holes**, and these holes can also move from one part of the crystal to another (though they cannot exist beyond the crystal). Because a hole in a crystal behaves like a positive charge, movement of holes also amounts to electric current, and in most metals when electric current flows, part of the current is due to hole movement and the rest to electron movement.

There are some materials for which the relative amount of electrons and holes can be adjusted. The materials we call **semiconductors** (typically silicon) can have tiny amounts of other materials added to the pure metal crystals during the refining process; this action is called **doping**. The result is that we can create a material that conducts mainly by hole movement (a P-type semiconductor) or one that conducts mainly by electron movement (an N-type semiconductor). In addition, the number of particles that are free to move (free particles) is less than in a metal, so that the movement of the particles is faster (for the same amount of current) than it is in metals, and the movement can be affected by the presence of charges (which attract or repel the electrons or holes and so interfere with movement). The movement can also be affected by magnets. Semiconductors are important because they allow us to control the movement of electrons and holes in crystals.

Summary

- All materials can be charged, for example by dislodging electrons. For every positively charged material there is a negatively charged material, because charging is caused by separating electrons from atoms, and the electrons will eventually return.
- The movement of electrons is what we detect as electric current, and the 'pushing force' for the current, caused by the attraction between positive and negative, is called voltage.
- Voltage is the cause of current.
- Materials can be roughly classed as conductors or insulators. Conductors have closely packed atoms which can share electrons, so that electron movement is easy, and electric current can flow even with only a small voltage. Insulators have more separation between atoms, electrons are not shared, and even a very high voltage will not cause any detectable current to flow. When we work with conductors we use low voltage levels and comparatively high currents. When we work with insulators we can use high voltage levels and very low currents.
- The third class of material is the semiconductor. Semiconductors can be natural, but better results are obtained by refining materials and deliberately adding impurities that will alter the number of free electrons or holes. The importance of semiconductors is that they make it possible to control the flow of charged particles, and this is the whole basis of modern electronics.

■ Note

Before semiconductors were discovered, electronic vacuum tubes (called **valves** in the UK) were used to control electron (not hole) flow. The principle, dating from about 1904, is that when electrons move in a vacuum, their movement through a wire gauze (or grid) can be controlled by altering the voltage on the grid. Vacuum tubes are still used where large voltages and currents have to be controlled, such as for high-power radio transmitters, and also for cathode ray tubes (for older television receivers and for the measuring instruments called oscilloscopes), but their use for other purposes has died out. Even the oscilloscope requirement is now being replaced.

■

Electromagnetic Waves

AC would be important enough even if it only provided a way to generate and distribute electrical power, but it has even greater importance. A hint of this came in 1873 when James Clerk Maxwell published a book containing equations that showed that an alternating voltage could generate waves of voltage and magnetism in space, and that these waves would travel at the same speed as light. He called these waves electromagnetic waves, and from there it was a short step to show that light was just one of these waves.

Why just one? These waves differ from each other in two ways. One is the number of waves that pass a fixed point per second, called the **frequency** of the waves. The other is the **amplitude** (the amount of rise and fall in each wave) (Figure 1.9). Amplitude determines the energy of the wave, so that a large amplitude of a light wave means a bright light. Frequency affects how easily a wave is launched into space and how we detect it.

Definition

Low-frequency electromagnetic waves are called **radio waves**, and we generate them and detect them nowadays using electronic methods. To put figures to these quantities, waves with frequencies below 100 kHz (one-hundred-thousand complete cycles per second) are classed as very low frequency (VLF) and are used mainly for time signals and for some long-distance communications. Waves of around 1 MHz (one-million hertz) frequency are called medium-wave, and a large number of entertainment radio transmitters use this range. Waves in the range 10 MHz to around 50 MHz are classed as short waves, used for communications, and the very high-frequency (VHF) range 50 MHz to 200 MHz is also used for similar purposes. As the frequency is increased, the range of useful communications along the Earth's surface decreases, but the range in space (as from a satellite to Earth) is much greater.

■ **Note**

The amplitude that is quoted for a wave is usually the peak amplitude. The figure of peak-to-peak amplitude is used when the wave is not symmetrical.

■

The range from 300 MHz to 1000 MHz is ultra-high frequency (UHF), used for television transmissions, and once we get to using the unit of GHz (1 gigahertz is equal to 1000 MHz)

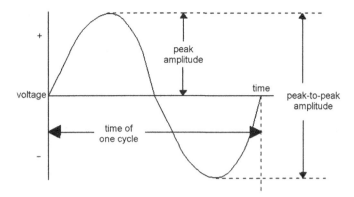

Figure 1.9:
Amplitude. The peak amplitude figure is used for symmetrical waves, like sinewaves. Peak-to-peak readings are used for waves whose shape is not symmetrical

then the signals are in the **microwave** ranges, used for mobile phones, satellite communications, and radar. These names are only rough indications of a range, and because we have found it necessary to use higher and higher frequencies over the years we have had to invent names for new ranges of frequencies that we once thought were unusable.

■ **Note**

Of all the possible microwave frequencies, only one (2.45 GHz) has a strong heating effect on anything that contains water. Other frequencies, such as are used for mobile phones and other communications (including satellite broadcasting), cause only negligible heating effects on materials unless very high powers are used.

■

One lower frequency range is particularly significant to us, and is called the **audio** range. This is the range of frequencies between about 30 Hz and 20 kHz, and its significance is that this is the range of frequencies of sound waves that we can hear (bats might have a different definition). A microphone, for example, used in a concert hall would provide an electrical output that would consist of waves in this range. For speech, we make use of a much smaller range, about 100–400 Hz.

■ **Note**

A **microphone** is an example of an important device called a **transducer**. A transducer converts one form of energy to another, and for electronics purposes, the important transducers are those that have an input or an output which is electrical, particularly if that input or output is in the form of a wave. For audio waves, the output transducer (converting electrical waves into sound waves) is a **loudspeaker**.

■

All the electronic methods that we know for generating waves are subject to some limitation at the highest frequencies, and methods other than electronic methods of generating and detecting waves have to be used. At around 1000 GHz, the waves are called **infrared**, and their effect on us (and any other objects) is a heating effect on all materials that absorb the waves, so that we can detect these waves coming from any warm object. We can also use electronic transducers to convert infrared signals into electrical signals. Higher frequencies, to about 100,000 GHz, correspond to the infrared radiation from red-hot objects, and one small range of frequencies between 100,000 and 1,000,000 GHz is what we call **light**. Higher still we have X-rays, gamma rays and others which we find very difficult to detect and cannot generate for ourselves.

As far as electronics is concerned, we make most use of the waves in the range from a few hertz to several tens of gigahertz, and radio technology is concerned with how these waves are generated, used to carry information, launched, and detected. Radio, in this respect, includes television and cellular telephones, because the use of the waves is the same; only the information is different.

Waveforms

The waveform of a wave is its shape, and because an electromagnetic wave is invisible the shape we refer to is the shape of the graph of (usually) voltage plotted against time. The instrument called the oscilloscope (see Chapter 17) will display waveforms; we do not need to draw graphs from voltage and time readings. We are seldom particularly interested in the shape of the waves that are transmitted through space, and waveforms are of interest mainly for waves that are transmitted along wires and other conductors in electronic circuits.

The simplest type of waveform is the shape of the wave that is generated by the magnet and coil arrangement that Faraday used (which was illustrated in Figure 1.6). The shape that this generates is called a **sinewave** (or sine wave) and it is the same shape as a graph of the sine of an angle plotted against the angle. What makes the sinewave particularly important is that any other shape of wave can be created by mixing sinewaves of different frequencies, and any shape of wave can be analyzed in terms of a mixture of sinewaves.

Though a pure sinewave is the simplest wave, its uses are confined to AC power generation and to radio transmitters. Most of the waves that we use in electronics are a long way from a sinewave shape, and one special type, the **pulse**, has become particularly important from the second half of the twentieth century onwards.

Take a look at the two shapes in Figure 1.10. Wave (a) is called a **square wave**, for obvious reasons, and it is used particularly when a wave is used for precise timing. Because the edges of the wave are sharp, each can be used for starting or stopping an action. If you used a sinewave there would be some uncertainty about where to start or stop, but the steep edge of

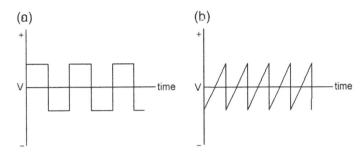

Figure 1.10:
Other waveforms: (a) square waves; (b) sawtooth or sweep waveform

the square wave makes the timing action much more certain. For example, the time needed to change the voltage level of such a wave might be only 50 nanoseconds or less (1 nanosecond is a thousandth of a millionth of a second).

The square wave is nothing like a sinewave, and it can be generated naturally by switching a steady voltage on and off rather than by any type of rotating generator. Very precise square waves are generated by electronic circuits that use a vibrating quartz crystal to control the frequency, and these circuits are used in clocks and watches. Even greater precision is obtained from the **atomic clock** that uses the natural vibration of atoms as a fixed frequency.

The other wave, in Figure 1.10(b), is called a **sweep** in the USA or a **sawtooth** in the UK. It features a long, even, rise (or fall) followed by a fast return to the starting voltage. Before 1936, a wave of this shape would have been an academic curiosity, but this is the shape of wave that is needed for a cathode ray tube, for television, or for radar, and we will look at it again in Chapters 8 and 17. The sawtooth is also an important waveform that is used in electronic measuring instruments.

What marks these waves out as totally different from the sinewave is that they show very sharp changes of voltage. A sinewave never changes abruptly; its rate of change is fixed by its frequency and amplitude. These square and sawtooth waves can change in a time that bears no relation to the frequency or the amplitude. A square wave might have a frequency of only 1 Hz, but change voltage in less than one-millionth of a second (a microsecond, written as 1 μs).

Summary

The waveform is the shape of an electrical wave, and the most fundamental waveform is the sinewave that is generated by a coil rotating between the poles of a magnet. The important features of a waveform are its frequency, the number of times a wave repeats per second, and its amplitude, the height of the wave. Sinewaves are used mainly in radio applications, but other waveshapes such as square and sawtooth waves are used, particularly where timing is important.

Pulses

The pulse is a waveform, but one that you might not recognize as a wave because the time of the pulse is very short compared to the time between pulses. Figure 1.11 illustrates three different pulse shapes, all of which share a very sharply rising portion, the **leading edge**. For a negative pulse, this leading edge would be a sharp fall in voltage.

Definition

A pulse is a rapid change in voltage which is of very short duration compared with the time between pulses.

For example, a pulse might repeat at a rate of 1 kHz, 1000 pulses per second. The actual pulse might have a duration, a pulse time, of only 10 μs, so that the change in voltage lasts only for 10 μs in the 1 ms (millisecond) between pulses. That makes the time of the pulse (its **duty cycle**) 1/100 of the time between pulses, and these are fairly typical figures. Pulses are used for timing, and they have the advantage that they use very little energy because the change in voltage is so short. A pulse can be used to start an action, to stop an action, or to maintain an action (such as keeping a wave in step, synchronized, with the pulses).

Modern digital electronics systems, particularly computers, rely heavily on the use of pulses, and when we work on these systems we are not greatly concerned about waveshapes, only about pulse **timing**. There are many things that can change a waveshape, making it very difficult to preserve the shape of a wave. By contrast, it is more difficult to upset pulse timing by any natural means, so that circuits which depend on pulse timing are more reliable in this respect than circuits that depend on waveshapes.

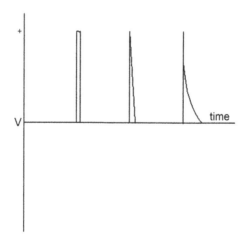

Figure 1.11:
Examples of pulse waveforms. The common factor is a sharp leading edge

The classic example is sound recording. Hi-fi systems in the past tried to work with a waveform that had a very small amplitude (the output from a gramophone pickup) and keep the shape of the waveform in the form of a copy that had a much larger amplitude. This large-scale copy was used to operate loudspeakers, and we called the whole exercise **amplification**.

The problem with this system is that a copy of the waveform is never perfect, something that is even more obvious when you make a copy of a copy. Any blemish on the surface of the record, any false movement of the stylus, any interfering signals in the circuits all will make the copied waveform inaccurate, a process we call distortion.

Nowadays the sound is recorded digitally as a set of pulses, using the pulses to represent numbers, and each point in a waveform is represented by a number. Using pulses for counting, we can ensure that the numbers are not changed, so that when the stream of numbers is converted back to a wave, the shape of the wave is exactly the same as was recorded. This is the basis of **compact discs**, the most familiar of the digital systems that have over the last few years replaced so many of the electronics methods that we grew up with. There will be more of all that in Chapters 11 and 12.

Actions on Pulses

There are two actions that can be carried out on pulses and on square-shaped waves that are important for many purposes. One of these is **differentiation**, and this can be achieved by passing a pulse or square wave into a circuit that selectively passes only the fast-changing part of the input. Figure 1.12 shows the result of differentiating, which converts the pulse or square wave into a pair of sharp spike shapes. These spikes are very short pulses, and we can use them for timing. We can select either a positive or a negative spike by using other circuits.

The other action is called **integration**, and it is a form of averaging or smoothing. A typical action is illustrated in Figure 1.13, showing a set of pulses as the input to an integrating circuit (an integrator). The output is a steady rise in voltage, and eventually this will become

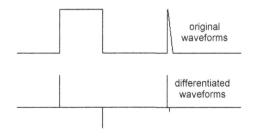

Figure 1.12:
Differentiating action, illustrated on two waveforms. The action emphasizes the sharply changing portions of the waves

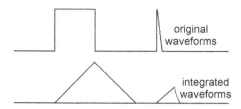

Figure 1.13:
Integrating action. The action smooths out sharp changes, altering a steep rise into a sloping rise, for example

steady at a value which is the peak voltage value of the pulses. This is the opposite action to differentiation, removing rapid changes from a waveform.

There is another waveform that is very important in almost all branches of electronics, the sawtooth or sweep wave. This is obtained by integrating part of a square wave, and was illustrated earlier in Figure 1.10(b). The steady rise (or fall) of voltage is called the sweep portion, and the rapid return (the portion that is not integrated) is called the **flyback**. We shall meet this type of wave again in connection with television, oscilloscopes, and digital voltmeters.

Summary

Pulse and square waveforms have sharp changes of voltage and can be used for timing. These waveforms can be differentiated or integrated by using suitable circuits. The action of differentiation emphasizes sharp changes in a wave; the opposite action of integration smooths out sharp changes.

Definition

Electronic circuits can be divided into **analog** and **digital** types. Analog circuits are used to operate on waves, preserving the shape of the wave. Digital circuits work with pulses and the shape is not important, only the timing.

Early applications for electronics required actions such as amplification, the creation of a copy of a wave with larger amplitude. The type of circuits used for these actions are classed as analog circuits. As the twentieth century progressed, circuits that counted using pulses became more important and such circuits are digital circuits. At present, digital circuit methods are steadily replacing the older analog methods, and this progress is reflected in this book.

CHAPTER 2

Passive Components

Electronic **components** are the building blocks of an electronic circuit, and all electronic circuits are created by joining components together. At one time this was done by soldering wires between the terminals of components, but nowadays the connections are more likely to be made using metal tracks on an insulating board (a **printed circuit board** or **PCB**). On a PCB holes are drilled into the metal tracks, so that components located on the insulated side can be attached and connected by pushing their connecting wires through the holes so that the wires can be soldered to the metal track. We will come back to that in Chapter 5. Often now both the connections and the components are all contained in a single piece of silicon, the **integrated circuit** (**IC**), which we shall look at in Chapter 3. In such a circuit, both steady and alternating voltages will exist together, and several types of components do not behave in the same way to alternating voltages as they do to steady voltages.

In addition, components can be **active** or **passive**. Active components are used to copy (**amplify**) waveforms and to switch voltages and currents on and off under electrical control. Such active components need an **input** signal (a waveform) to control an **output** signal, and they also need some source of power, which is usually a steady voltage supply. A circuit that contains active components can produce an output waveform which provides more power (voltage multiplied by current) than its input waveform. In other words, active components can provide **amplification** of power.

■ Note

There are passive components, such as transformers, that can provide amplification of voltage (but at reduced current) or amplification of current (at reduced voltage), but *not* amplification of power.

■

Passive components always *reduce* the power of an input waveform, so that an output wave from a circuit that contains only passive components is always at a lower power than (or the same power as) the input. Passive components do not need any additional steady voltage supply to enable them to deal with waveforms. A complete electronic circuit will normally consist of both active and passive components, arranged so that the passive components control the action of the active components and act as a path for signal waveforms. Take a look now at the most common passive components, resistors, capacitors, and inductors.

Electronics Simplified. DOI: 10.1016/B978-0-08-097063-9.10002-0

Resistors

Resistors are the most common of passive components. We saw in Chapter 1 that the quantity called resistance connects current and voltage, so that in any circuit or part of a circuit, the ratio voltage/current (*V/I*) is the resistance. If we measure the voltage across a resistor in volts (V) and the current through the resistor in amps (A), then the resistance is in units of ohms (Ω). Using the Greek letter omega for ohms avoids the confusion that would be inevitable if we used a capital letter O.

Definition

A resistor is used to control current or to convert a wave of current into a wave of voltage, using the $I = V/R$ or $V = RI$ relationship.

Every part of a circuit has some resistance to the flow of current, and when we specify a resistor as a circuit component we mean a component that is manufactured to some precise (or reasonably precise) value of resistance and used to control the amount of current flowing in some part of a circuit. Though it is possible to manufacture resistors with low values of a fraction of an ohm, most of the resistors that we use in electronics circuits have higher values of resistance, and to avoid having to write values like 15,000 Ω or 2,200,000 Ω we use the letter k to mean 'thousand' and M to mean 'million' and we omit the omega sign. In addition, the letter R is often used to mean ohms, because typewriters (which, unlike word-processors, do not have the omega symbol) are still being used. Another way of making values clearer is to use the letters R, k or M in place of a decimal point, because decimal points often disappear when a page is photocopied.

For example, using these conventions, you would write 15,000 Ω as 15k and 2,200,000 Ω as 2M2. A resistance of 1.5 Ω would be written as 1R5, and a resistance of 0.47 Ω would be written as 0R47. The small letter m can be used to mean milli, so that 4.7 mΩ means 0.0047 ohms.

Resistors can be manufactured to practically any value that you want, but in practice there is no point in having a vast range. A standard set of **preferred values** is used, and this also fits in with the manufacturing tolerances for resistors and other components. For example, the standard values of 1.0 and 1.5 allow manufacturing with 20% tolerance with no rejected resistors. This is because if you take any pair of values on the scale, then a value which is 20% high for one value will overlap the amount which is 20% low for the next value. For example, 20% up on 1.0 is $1.0 + 0.20 = 1.20$ and 20 down on 1.5 is $1.5 - 0.3 = 1.2$, so that these values can overlap: a resistor of value 1R2 could be a 1R0 which was on the high side or a 1R5 which was on the low side. We can pick numbers from the standard set (for 5% tolerance) to suit 10% or 20% tolerances, as Table 2.1 shows. Resistors with 20% tolerance are hardly ever used nowadays because modern electronics demands more precise values.

The preferred value numbers need only be in the range shown here, because we can multiply or divide by 10 to obtain other ranges. For example, the number 4.7 can be used

Table 2.1: Preferred values for
5%, 10%, and 20% tolerance.

5%	10%	20%
1.0	1	1
1.1		
1.2	1.2	
1.3		
1.5	1.5	1.5
1.6		
1.8	1.8	
2		
2.2	2.2	2.2
2.4		
2.7	2.7	
3		
3.3	3.3	3.3
3.6		
3.9	3.9	
4.3		
4.7	4.7	4.7
5.1		
5.6	5.6	
6.2		
6.8	6.8	6.8
7.5		
8.2	8.2	
9.1		

in the form 4R7 for 4.7 Ω, or as 0R47 (0.47 Ω), 47R, 470R, 4k7, 47k, 470k, 4M7, 47M, and so on.

■ Note

Resistor values are indicated on the body of resistors using a color code, though on sub-miniature components the value is often printed in alphanumeric characters (and may need a magnifying glass to read). See Appendix B for details of the color code, and websites that show the shape of typical components.

■

Summary

Resistors are manufactured using a range of preferred values that ensures there will be no rejects. Tolerances of 1% and 5% are commonly used, and closer tolerances can be obtained. Values are often written using the letter R in place of the ohm sign or the decimal point, and using k (kilo) to mean thousand and M (mega) to mean million. This allows a value to be specified without the need to write a large number of zeros.

A resistor behaves in the same way to all voltages and currents, whether these are steady voltages (or currents) or waveform voltages (or currents). In other words, the relationship $V = RI$ is always true for a resistor that is kept at a constant temperature (Ohm's law). When a current passes through a resistor there will be a voltage across the resistor, and power is converted into heat (Figure 2.1). Just as the $V = RI$ relation can be written in three ways, the power equations also exist in three forms.

Wherever there is resistance in a circuit, electric power is converted into heat, and this represents **dissipation**, waste, loss of energy. In addition, the conversion of electrical energy into heat means that the temperature of a resistor will rise when it is passing current. Unless the resistor can pass on this heat to the air it will overheat and be damaged. Figure 2.1 shows the relationship between voltage, current, resistance, and lost power, and also shows the symbols that are used to represent a resistor on a circuit diagram. Both the block and the zigzag symbols are in use, but in Europe the block is the preferred symbol nowadays.

This power loss (dissipation) is always associated with resistance (whether of resistors or in other components). If the dissipation is large some method has to be used to keep the components from overheating, and this usually takes the form of cooling fins so that the heat can be more efficiently transferred to the air. There are some components (such as capacitors, see later) that do not dissipate any measurable amount of heat, but any circuit will inevitably contain some resistors whose heat will spread to other components. Heat dissipation is particularly important in a circuit that contains active components, as we shall see in Chapter 3.

When a resistor carries a steady current, there will be a steady voltage across the resistor ($V = RI$); and when there is a steady voltage placed across a resistor, there will be a steady

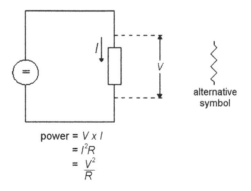

power = $V \times I$
$= I^2 R$
$= \dfrac{V^2}{R}$

Figure 2.1:
Power dissipated by a resistor. There are three versions of the formula, so that you can use whichever is most suitable. For example, if you know values of R and I, use the RI^2 formula

current ($I = V/R$) through the resistor. The relationship applies also to waves, so that if a waveform of current passes through a resistor there will be a voltage waveform across the resistor whose value can be found from $V = RI$. When a resistor is used like this to obtain a voltage wave from a current wave, we call it a **load resistor**. Load resistors are used along with **active** components.

■ Note

When we use these $V = RI$ relationships with alternating currents and voltages, we normally use peak values for both voltage and current. Another option is to use root mean square (RMS) values (see Chapter 4) for both quantities. Whatever type of measurement you use must be used consistently: you cannot multiply a peak value of voltage by an RMS value of current and get anything useful.

■

Resistors can be used to reduce the amount (amplitude) of a signal. Suppose, for example, that we connect two resistors in series (one connected to the end of another) as shown in Figure 2.2. If a signal voltage is connected across both resistors, as illustrated, then the output across just one single resistor is a smaller signal voltage, and the size of this signal can be calculated from the sizes of the resistors. As a formula, this is:

$$V_{\text{out}} = V_{\text{in}} \frac{R_y}{(R_y + R_x)}$$

and it allows us to adjust signals to whatever amplitude we want to use. Suppose, for example, that R_y in Figure 2.2 is 10k and R_x is 15k, and we have a 20 V signal at the input. Since the total resistance is 25k, the output is $20 \times 10/25$, which is 8 V.

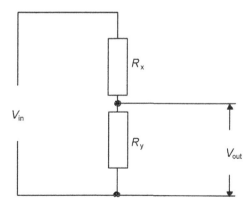

Figure 2.2:
The potentiometer or voltage divider circuit. The output voltage is a fraction of the input which can be calculated if the resistor values are known

■ Note

The dot that is placed where lines join in this **circuit diagram** is a way of emphasizing that these lines are electrically connected. When you see lines crossing, with no dot, on a modern circuit diagram this means that there is no connection between the lines. Older diagrams used a hump to indicate a line crossing. We will look in more detail at circuit diagrams later.

■

The combination of two resistors (Figure 2.2) is called a **potentiometer** or an **attenuator**, and we can manufacture a component, an adjustable (**variable**) potentiometer which allows us to vary the values of both resistors together, keeping the total resistance constant. The symbol is shown in Figure 2.3, and you can think of it as a resistor with an extra contact that can be moved in either direction. This allows the output voltage to be adjusted (by altering the position of the contact) from the maximum (which is the same as the input) to zero. The potentiometer can be used, for example, as a volume control in a radio.

■ Note

The resistors in Figure 2.2 are connected in **series**, meaning that the current must pass through both resistors equally, one after the other, making the voltages across each resistor different unless the resistance values are the same. The alternative is **parallel** connection, in which the current splits between the resistors. When components are connected in parallel, the voltage across each of them is the same but the currents through each will be different unless the resistance values are the same. You will see several examples of series and of parallel connections in the course of this book.

■

Summary

The main use of resistors is to control current or as load resistors to convert current waves into voltage waves. The relationships $V = RI$, $I = V/R$ and $R = V/I$ hold for either steady or alternating voltages and currents provided the same types of units are used. Two resistors in series can be used to reduce (**attenuate**) a signal voltage, and a variable version of this arrangement is a potentiometer, used to adjust signal levels.

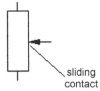

Figure 2.3:
The variable potentiometer symbol. This component is used to provide an adjustable voltage division and a typical use is as a volume control

Capacitors

A capacitor is a gap in a circuit, a sandwich of insulating material between two conductors which has **capacitance** (see later). As far as DC is concerned the capacitor is a break in the circuit, but a capacitor will allow AC to pass, so that it allows us to separate AC from DC. The symbol for a capacitor (Figure 2.4) shows it as two conducting plates with a gap between them, and this can be used as a way of manufacturing capacitors, though we more usually find a solid insulator between the plates.

Two exceptions are variable capacitors and electrolytic capacitors. Variable capacitors use two sets of plates that mesh with each other (not touching), and because one set of plates is carried on a spindle, turning the spindle will alter how much the plates mesh, and so alter the (small) capacitance between them. This has been used widely in the past for tuning radios. The electrolytic capacitor, by contrast, uses an acid jelly held between metal plates, and the insulator is hydrogen gas that is generated by chemical action. This type of capacitor is used when a very large value of capacitance is needed in a small volume.

As far as steady voltages or currents are concerned, the capacitor is just a gap, a break in a circuit so that no steady current can flow. When you place a steady voltage across a capacitor there is no **steady** current (but there can be a momentary current, as we shall see later). It is a different matter when an alternating voltage is used. When you move electrons on to one plate of a capacitor, the same number of electrons will leave the other plate, because of the electrostatic effect of like charges repelling each other. When you move electrons alternately to and from one plate, the same waveform will occur on the other plate, just as if it had been connected through a circuit. You can measure the alternating voltage across the capacitor and the alternating current through it and find the size of quantity which is given by V/I. This quantity is called **capacitive reactance**, and given the symbol X_C. This has units of ohms (because it is a ratio of voltage to current), but it is not the same as resistance. We will come back to that point later.

Unlike a resistor, a capacitor does not have a fixed value of this reactance quantity, because if you change the frequency of the supply wave, the reactance of a capacitor will change. When frequency increases, reactance decreases, and when frequency is decreased, reactance increases. There is, however, a quantity called **capacitance** which depends on the physical

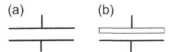

Figure 2.4:
The capacitor symbol. The basic symbol is this parallel-plate type (a) that indicates the nature of a capacitor as a pair of conductors separated by an insulator. (b) The symbol for an electrolytic capacitor, used for large capacitance values

measurements of the capacitor and the type of insulating material between the plates, not on the frequency of the voltage waveform. The value of capacitive reactance can then be calculated if you know values for capacitance and for the frequency of the alternating voltage.

Definition

When a charge, amount Q, is placed on the plates of a capacitor, there will be a voltage V between the plates. The capacitance C is defined as $C = Q/V$, with charge measured in coulombs (Q) and V in volts, while the unit of capacitance is farads. This, however, is not a practical way of measuring capacitance because it is difficult to measure charge precisely.

The natural unit for capacitance is the coulomb per volt, called the farad (named after Michael Faraday), but this unit is too large for most of the sizes of capacitors that we use for electronics circuits. We therefore use the smaller units of microfarad (μF), nanofarad (nF), and picofarad (pF). A microfarad is one-millionth of a farad, the nanofarad is one-thousandth of microfarad, and the picofarad is one-millionth of a microfarad. For example, a variable capacitor might have a maximum value of 300 pF; an electrolytic capacitor might have a value of 5000 μF. Capacitors of 1 F or more can be manufactured for use as backup supplies in low-consumption electronics circuits.

Summary

A capacitor consists of an insulator between conducting plates, and does not pass steady current. Alternating voltages will cause an alternating current to pass through a capacitor, and the ratio of V/I is constant if the frequency is not changed. This ratio is called capacitive reactance, measured in ohms. A more fundamental quantity, called capacitance, can be calculated, in units of farads, from the dimensions of a capacitor and the type of insulator, and this quantity is a constant for a capacitor. The reactance at any frequency can be calculated from the capacitance value.

When an alternating voltage is applied to a capacitor and alternating current flows, the voltage wave is not in step with the current wave, but occurs one-quarter of a wave later (Figure 2.5). Contrast this with the behavior of a resistor, which is also illustrated in this diagram. Because the maximum current through a capacitor happens at the time of zero voltage, and the maximum voltage happens at the time of zero current, there is no power dissipation from a perfect capacitor (one which has no resistance). Nothing is perfect, but capacitors get pretty close in this respect and their dissipation is usually almost immeasurably small.

The amount by which voltage and current are out of step is expressed as a **phase angle**. If you think of a cycle of a wave as being caused by a coil rotating between the poles of a magnet, one complete wave corresponds to one complete turn of the coil, turning through $360°$. On this basis, half a wave corresponds to $180°$ and quarter of a wave to $90°$. We say, then, that the capacitor causes a $90°$ **phase shift** between voltage and current for an alternating supply, with

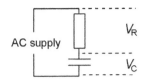

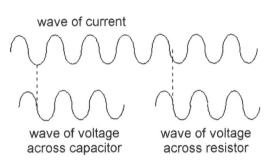

Figure 2.5:
Phase angle. The phase of voltage across the capacitor is a quarter of a wave time (90°) later than the phase of current (which is also the phase of the voltage across the resistor)

the current wave ahead of the voltage wave. We say that in the capacitor, current **leads** voltage or, looking at it the other way round, that voltage **lags** current.

■ Note

This idea of phase is very important in all branches of electronics, and you will need to recall it when we look at stereo radio broadcasting and color television principles. Phase is a quantity that we can alter for a wave just as we can alter frequency or amplitude, and when we use the idea of phase we are always comparing one wave with another. So far, we have been comparing a current wave with a voltage wave at the same point in a circuit, but you could equally compare the phase of one voltage wave with another, or compare the phase of a wave at one point in time with the phase it had earlier.

■

When a circuit contains a capacitor the wave of voltage in that complete circuit will not be in step with the wave of current. Suppose, for example, that a circuit contains both a capacitor and a resistor. The wave of voltage across the resistor will be in step with the current wave, but the wave of voltage across the capacitor will not be in step with the current wave through the capacitor. The result is that the total voltage across the circuit cannot be calculated simply.

For example, in the circuit of Figure 2.6, the total voltage V_T is not equal to $V_C + V_R$. There are ways of calculating this, but they are a long way from simple addition, and this book is not concerned with mathematics. Another consequence of this is that the power dissipated in the resistor is no longer found by multiplying voltage and current, because the peak voltage does

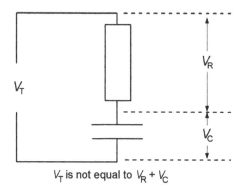

V_T is not equal to $V_R + V_C$

Figure 2.6:
One effect of phase shift is that the total voltage across a series R and C circuit is not equal to the sum of the separate voltages across the components

not occur at the same time as the peak current (because of the phase shift). To be mathematical for a moment, the power has to be found from the equation:

$$P = V \times I \times \cos \Phi$$

where Φ is the phase angle between voltage and current. If Φ is 90 then $\cos \Phi = 0$, and the power dissipated is also zero.

■ Note

The cosine (or **cos**) of an angle is a quantity that varies between 0 and 1, depending on the size of the angle. A graph of cosine of angle plotted against angle from 0° to 360° is a waveform shape of graph (a sinewave shifted by 90°).

■

Summary

Alternating current through a capacitor is one-quarter of a wave, 90°, ahead of the wave of voltage. This has two effects. One is that there is no dissipation in a capacitor except from the resistance of its conductors. The other is that the presence of a capacitor in a circuit causes the waves of voltage and of current in the whole circuit to be out of step.

Charge and Discharge

Capacitors act as insulators for steady voltages, and as reactances for waves, but their behavior both with steady voltages and with changing voltages such as pulses is also important. The capacitor can accumulate a tiny amount of electrical charge when it is connected to a steady voltage, but if the connection is made (as it almost always is) through a resistor, then this

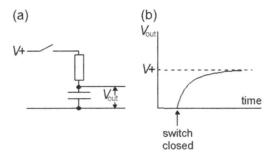

(a)　　　　　　(b)

Figure 2.7:
Charging a capacitor through a resistor. In this circuit (a), the voltage across the capacitor rises with a rate that is not constant, giving a curved graph (b). This is an exponential increase

charging action takes time, even if this time is measured in microseconds, and the time is not a simple measurable quantity.

Look, for example, at what happens in the circuit shown in Figure 2.7(a). This shows a voltage supply with a switch, a resistor, and a capacitor. Starting with the switch open, so that the circuit is disconnected, there will be no voltage across the capacitor. When the switch is closed, current flows momentarily and it will charge the capacitor, but as the voltage across the capacitor increases the amount of current is reduced (if you think in terms of water pouring through a pipe into a jug it has a shorter distance to fall), so that the rate of charging slows down.

The result is that a graph of voltage across the capacitor plotted against time takes the form shown in Figure 2.7(b). This is the type of graph shape that is called an **exponential rise**, and what makes it interesting is that it is, in theory, never totally complete. If we have a charged capacitor and we connect a resistor across its terminals, the voltage plotted against time will give the graph shape called an **exponential decrease**. Mathematically, the graph is described using a universal constant called 'e', the **exponential constant** which also appears in calculations of things you might think were not related, such as compound interest, population growth, or the decay of radioactivity. A convenient rule of thumb for capacitor and resistor combinations makes use of what is called a **time constant**.

Definition

The time constant of a combination of a capacitor and a resistor is the value of capacitance multiplied by the value of resistance. If the capacitance value is in units of farads and the resistance is in units of ohms, the time constant R × C is in units of seconds. More practical units are kΩ for the resistor and nF for the capacitor, giving time in μs. For example, using a capacitor of 20 nF and a resistor of 100k gives a time constant of 2000 μs, which is 2 ms (milliseconds).

The importance of the time constant is that we can take it that charging or discharging is over for all practical purposes (meaning about 95%) after a time of three time constants. For more precision, a value of four time constants can be used, but we will stay with the value of three times in this book. This makes it easy to work out the times for the waveform that is produced when a resistor and capacitor are used in a charging or a discharging circuit. Suppose that a resistor is connected across a charged capacitor, as in Figure 2.8. The shape of the voltage/ time graph is then as shown in the drawing, once again taking the process as being complete after three (or four) time constants.

For example, if the input to the circuit of Figure 2.9 is a square wave whose flat portion takes more than three time constants, we can draw the output waveform fairly easily. The first part is a charging curve taking three time constants, and the last part is a discharging portion that also takes three time constants. If the top of the square wave had a duration of more than six time constants, the remainder is unaffected. The effect is one that we noted in Chapter 1, of **integration** of the square wave.

Figure 2.10 shows a slightly altered circuit in which the components are rearranged and the output voltage is across the resistor. Now when there is a sudden rise of voltage at the input, the capacitor has no time to charge, and the voltage across it is zero, which means that all of the input voltage appears across the resistor. Then as the capacitor charges, the voltage across the resistor drops to zero in the time of three time constants. If the input is then suddenly taken back to zero the process repeats, with the voltage across the resistor dropping (so that the capacitor maintains its change), and then reducing to zero in three time constants. This is the action of a **differentiating** circuit.

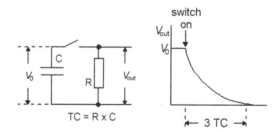

Figure 2.8:
Discharging a capacitor through a resistor. The graph shows an exponential decrease, and we can take it as being complete in three time constants

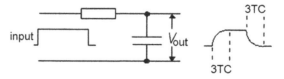

Figure 2.9:
The effect of this type of circuit on a square wave, showing the time in terms of time constant t. This is an integrating circuit

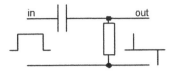

Figure 2.10:
The differentiating form of the circuit. Remember that the voltage across the capacitor cannot change instantly, so that when the voltage changes suddenly on one plate it must make the same change on the other plate, after which charging or discharging will alter the voltage

Inductors

An **inductor** is a coil of wire (or any other shape of conductor wound in a circle). Since this wire has resistance, an inductor will pass steady current when there is a steady voltage across it, but the fact that the wire is wound into a coil makes it behave as more than just a resistance for alternating current. If the resistance is low, we find that the alternating current through the coil lags almost 90° behind the alternating voltage across the coil, as illustrated in Figure 2.11, which also shows the symbol for an inductor.

For any inductor, the value of **reactance** can be measured, an **inductive reactance**, equal to alternating voltage divided by alternating current, using the units of ohms for reactance. This quantity is constant only for a fixed frequency. If you increase the frequency of the alternating voltage, the reactance of the inductor also rises.

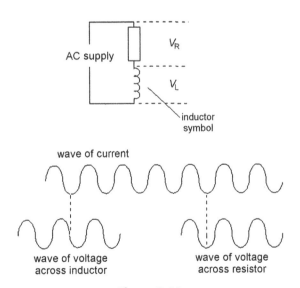

Figure 2.11:
Phase angles in a circuit containing an inductor and a resistor. The phase of voltage across the inductor (assuming it has no resistance) is 90° *ahead* of the phase of current (which is also the phase of the voltage across the resistor)

Once again, there is a quantity, called **inductance**, that can be calculated from the dimensions of the coil and knowledge about its magnetic core (if any core is used). This quantity of inductance is constant for a coil. The more turns there are on the coil, the greater the inductance, and the inductance is also (greatly) increased when the coil is wound on a magnetic material (a **core**). Inductance is measured in units called henry (H), named after the American pioneer Joseph Henry. We often use the small units of millihenries (mH) and microhenries (μH), and the abbreviation letter for an inductor is L.

Inductors are far from perfect: they have resistance because they are made from wire, so that there is always some dissipation from a coil, and the phase shift is never exactly 90°. Most of the coils that we use, however, have much larger values of reactance than of resistance, and the imperfections are not too important. Modern electronic circuits avoid the use of inductors as far as possible, but when they are used their symbol reminds you that an inductor is a coil.

■ Note

Some applications call for very small inductance values. One example is a resonant circuit (see later) for a television ultra-high-frequency tuner or for a satellite receiver. To make the very small values of inductance the wire does not even need to be coiled, and it is more usual to see a flat strip of metal used in place of a coil.

■

Summary

A coil of wire is simply a resistor as far as steady voltage is concerned, but for alternating voltages it behaves as an inductor. An inductor has inductive reactance, and causes a phase angle of almost 90° between current and voltage, with the voltage wave leading the current wave. The resistance of the wire means that an inductor is never perfect, but at the higher frequency ranges, the reactance can be very much greater than the resistance. The inductance of a coil can be calculated from its dimensions, and used to find reactance at any frequency. Reactance increases as frequency is increased. The dissipation in an inductor is only the amount you would expect from the resistance of its conductors.

■ Note

Why does a coil behave so differently to AC? The reason is that the AC generates changing magnetism around the coil, and the changing magnetism generates an alternating voltage that is out of phase, opposing the voltage that drives the alternating current through the coil. The net effect is to oppose current just as a resistor opposes current (though for a different reason).

■

■ **Note**

The combination of a resistor and an inductor also forms a **time constant**, but in practice this combination is seldom used in the way that we use resistor–capacitor time constants.

■

Transformers

The simplest transformer consists of two coils of insulated wire wound over the same core of magnetic material, one coil for input, the other for output. This has no effect as far as steady voltages are concerned, but for an alternating voltage the effect is very useful. When an alternating voltage is applied to one of the two coils, called the **primary**, this will generate an alternating voltage at the same frequency across the other coil, the **secondary**.

The ratio of these voltages is, ideally, the same as the ratio of the number of turns in the coils. For example, if the secondary coil has half the number of turns that the primary coil has, then the AC voltage across it will be half of the AC voltage across the primary. This arrangement is a **step-down** transformer, and you can just as easily make a **step-up** transformer, for which the voltage across the secondary coil is greater than the voltage across the primary. Figure 2.12 shows the circuit symbol and an elementary form of construction for a small transformer.

■ **Note**

What is happening depends on two effects. One is that the alternating current flowing through the primary winding creates alternating magnetism of the core, just as a steady current would create a steady magnetism. The other effect is that the alternating magnetism will generate an alternating voltage in the secondary coil, but a steady magnetism would not generate any voltage. The core acts as a magnetic link between the windings.

■

There is no gain of **power** in a transformer; it is a passive component and there will always be a loss of power. If you have 100 V across the primary coil and 1 A flowing, and a secondary that gives 200 V, then the secondary current cannot exceed 0.5 A. That is assuming no resistance and therefore no losses. We can come reasonably close to this perfection in very large transformers, such as are used on the National Grid, but not on small transformers unless they are being used at frequencies much higher than the normal 50 Hz or 60 Hz (USA) of the power mains. You can always expect loss of power in any small transformer, and this is indicated by a rise in temperature.

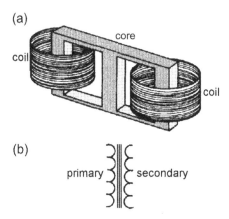

(a)

coil

core

coil

(b)

primary   secondary

Figure 2.12:
The transformer: (a) simplest practical arrangement; (b) symbol

The transformer was invented by Michael Faraday, and it is the main reason for our use of **AC** for electricity distribution. Transformers allow us to convert one alternating voltage into another with only very small losses (caused by the resistance of the wire in the coils). We can generate electricity at a voltage which is convenient, such as 25 kV (25,000 V), and convert this to 250 kV or more for transmission, because for the same power level, the current flowing in the cables will be one-tenth of the current from the generator. The less the current, the lower the power that is dissipated.

This is why we use pylons for electricity transmission, because we need to keep a large distance between cables at 250 kV (or more) and the earth. This is also why burying cables is ruinously expensive, because even if you can insulate the cables adequately, the cable will act as one plate of a capacitor (the earth being the other) and the alternating current will flow between the cable and earth. Since this is not exactly a perfect capacitor (the earth is moist) there will be heavy losses. If it is essential to bury cables, we need to use DC in the cables, adding more complication. It also adds another set of losses of valuable energy. Some of the proposals we see for dealing with our electricity generation would end up with us living in dark (but very green) caves.

The main use of transformers in electronics is in converting the 240 V AC mains supply to the low DC voltage that is needed for electronics in the power supply unit (PSU). The transformer converts the 240 V AC of the mains into a suitable lower voltage, and other components then convert this low-voltage AC into low-voltage DC.

Summary

When two coils are wound on the same magnetic core the result is a transformer. A transformer can change voltage and current levels with almost no loss of power. The ratio of (AC) voltages for a transformer is, ideally, the same as the ratio of the number of turns in the windings.

Resonance

We have seen that both capacitors and inductors affect the phase angle between current and voltage for AC. The effect, however, is in opposite directions, and a useful way to remember this is the word C-I-V-I-L. Say this as 'C − I before V; V before I for L' to remind you that for a capacitor (C) the current (I) wave comes before the voltage (V) wave, but the voltage wave comes before the current wave in an inductor (L). The US version of this is E-L-I-I-C-E, using E for voltage.

Suppose a series circuit contains both capacitance and inductance along with the inevitable resistance such as in Figure 2.13. How does such a circuit respond to alternating voltages? We can, of course, rule out any possibility of steady current, because the capacitor will act like a break in the circuit as far as DC is concerned. The interesting thing is that if we pass an alternating current, the voltage across the capacitor will be in opposite phase, 180°, to the voltage across the inductor (because, compared to the current, the voltage across the inductor is 90° leading and the voltage across the capacitor is 90° lagging) (Figure 2.14). The total voltage across the reactive components is the *difference* between the voltage across the inductor and the voltage across the capacitor.

This becomes particularly interesting when the reactance of the inductor is exactly the same size as the reactance of the capacitor. When this is true, as it must be at some frequency, then the sum of the reactances, in opposite directions, will be zero, and all that is left is the resistance due to the resistor, which can be quite small. This condition is called **resonance**, and in this **series resonant circuit** the current is maximum (equal to *V/R*, where *R* is the circuit resistance) at resonance. A graph of current plotted against frequency, near the frequency of resonance, looks as in Figure 2.15.

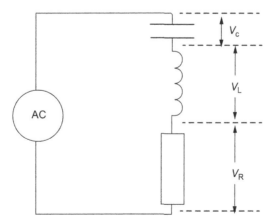

Figure 2.13:
A circuit containing capacitance, inductance, and resistance in series.
This is a series resonant circuit

Definition

A resonant circuit is one in which the effects of capacitance and inductance cancel each other out for one particular frequency.

There is another way of connecting a capacitor and an inductor (with the inevitable resistance) (Figure 2.16). This is a **parallel connection**, and in this circuit, DC *can* pass because the coil is a wire connection. If we apply only an alternating current supply, however, we find that this time the alternating **voltage** (rather than the current) becomes a **maximum** (not a minimum) at the frequency of resonance when the reactances are equal in size.

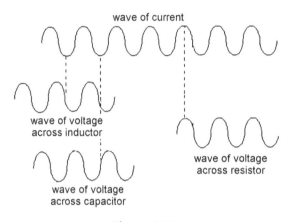

Figure 2.14:
How the voltages across the capacitor and across the inductor oppose each other. These voltages, for a perfect inductor, cancel each other out exactly at the frequency of resonance

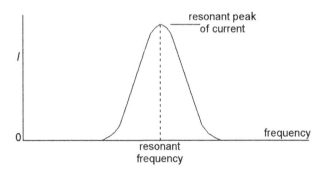

Figure 2.15:
How the current varies with the frequency for a series resonant circuit. This allows a particular frequency to be selected, but the series circuit is less common for practical uses than its counterpart, the parallel resonant circuit

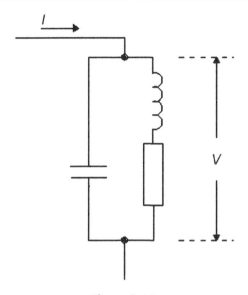

Figure 2.16:
The parallel resonant circuit. The voltage *V* across this circuit is a maximum at the resonant frequency

A resonant circuit can act like a selective transformer, delivering an output which is at a much larger voltage or current than the input, for one particular frequency (in practice, a small range or **band** of frequencies centered around one frequency). This is the effect that allows a radio or television to be tuned to one of a set of transmitting stations, using the selective transformer effect of the resonant circuit.

Resonant circuits are also important for timing and for transmitting signals. A piece of **quartz**, made in the form of a capacitor that uses the quartz as an insulator, will behave like a resonant circuit, with a step-up ratio at resonance that is much higher than can be achieved by any combination of inductor and capacitor, so that these quartz crystals are used to control the frequency of transmitters and also in clocks and watches.

Summary

When an inductor and a capacitor are used in the same circuit, their phase shifts are in opposite directions. When the sizes of the reactances are equal, the effects cancel so that for alternating signals, the only effect is of resistance. For a series circuit, this causes the current to be a maximum at the resonant frequency and for a parallel circuit the voltage is a maximum at resonance. The resonance effect is used for selecting a frequency or a small range (a **narrow band**) of frequencies for purposes such as radio tuning. A quartz crystal can be made to resonate, and is more efficient than any inductor/capacitor combination, so that quartz crystals are widely used for timing and frequency setting.

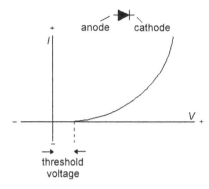

Figure 2.17:
The diode and its graph of current plotted against voltage. Even in the forward conduction direction, there is no current flowing when the voltage is small (typically 0.4 V for diodes constructed from silicon). There is no current in the reverse direction unless a very large reverse voltage is applied, which will cause the diode to break down (becoming open circuit)

Diodes

A diode is a passive component, but its construction follows the methods that are used for semiconductors (which are active components). A diode can be used with either steady or alternating supplies, but all resemblance to a resistor ends there, because a diode is not ohmic (see Chapter 1). A diode passes current in one direction only, and this is indicated by an arrowhead on the symbol that is used (Figure 2.17). The diode terminals are named **anode** and **cathode**, and in normal use current passes only when the anode is at a higher positive voltage than the cathode.

This illustration also shows a typical graph of current plotted against voltage. Unlike the corresponding graph for a resistor, this graph shows both positive and negative scales for current and voltage, because this allows us to show that the diode conducts in one direction only, and that it is not ohmic. The graph line is not straight even when the diode is conducting, so that there is no single figure of resistance that can be used; you cannot specify a 3k3 diode, for example. The resistance is very high when the current is low, and becomes lower as current is increased. Even when the voltage across a diode is in the **forward** (conducting) direction, the current is almost undetectable until the voltage has reached a threshold level of around 0.56 V for a silicon diode.

■ Note

A diode will **break down**, making it useless, if a large enough reverse voltage is applied, allowing excessive current to flow in the reverse direction. Diodes can be manufactured whose **reverse breakdown voltage** is precise and stable, and these **Zener diodes** (see later in this chapter) are used for providing a stable voltage level.

■

The effect of a diode on an alternating voltage supply is illustrated in Figure 2.18. The effect is like that of a commutator, allowing only half of the waveform to appear at the output. This effect is used in converting AC into DC, and also for a task called **demodulation** of radio waves (see Chapter 6). The circuit illustrates the simplest conversion circuit, half-wave rectification, and the much more common **full-wave bridge** circuit. A diode is a passive component, though the methods that are used to manufacture diodes are also used to manufacture active components.

Figure 2.19 shows a typical simple PSU which uses a diode rectifier bridge circuit along with electrolytic capacitors and a voltage stabilizer IC. The transformer supplies AC at a suitable (low) voltage and the output of the transformer is connected to the input of the diode bridge circuit. The input to the stabilizer is a voltage that is higher than we need at the output, typically +18 V for a 12 V output. The stabilizer contains a voltage reference source, a Zener diode that is operated with reverse voltage and which has broken down. Such a diode has a constant voltage across it even if the current varies, and it can be used in a circuit which compares this steady voltage with the output voltage of the chip, using this difference to control the output voltage.

■ Note

The power supplies for computers and other devices that use large currents (typically 20 A or more) at low voltage levels (typically 5 V or less) are constructed differently, using what is called a **switch-mode** power supply. This uses an oscillator (see later) to

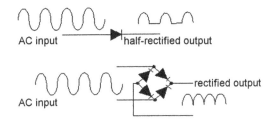

Figure 2.18:
Rectification of AC using diodes. The simple half-wave action is used for demodulation (see later), but the four-diode bridge circuit is almost universally used for AC to DC conversion for power supplies

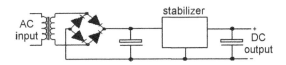

Figure 2.19:
A stabilized power supply, using a transformer, a diode bridge, and a stabilizer chip. The electrolytic capacitors ensure that enough charge is stored to maintain voltage for the short intervals when the output from the diodes approaches zero twice in each cycle

generate pulses at a high frequency, and these pulses are rectified and smoothed to provide the output. The output voltage is also used to control the action of the oscillator so that the output voltage remains constant even when the amount of current changes rapidly.

Other Diode Types

Zener diodes, as we have seen, are used with reverse bias, making use of the breakdown that occurs across a silicon diode when the reverse voltage is comparatively large. Breakdown occurs at low voltages (below 6 V) when the silicon of the diode is very strongly doped (mixed) with other elements, and such breakdown is termed Zener breakdown, from Clarence Zener who discovered the effect. For such a true Zener diode, the reverse characteristic is as shown in Figure 2.20(a). Another breakdown effect, **avalanche breakdown**, occurs in silicon with less doping and at higher reverse voltages.

Both types of diodes are, however, known as Zener diodes and those with breakdown voltages in the range of 4–6 V can combine both effects. The stabilization of a diode is measured by a quantity called dynamic resistance, which can be as low as 4 Ω. This quantity measures how effective the diode is in reducing small changes of voltage across it; the lower the dynamic resistance the better the stabilization.

A typical simple circuit is illustrated in Figure 2.21. The Zener diode is connected in series with a resistor which is used to limit the current (to avoid damaging the diode). If the supply voltage input varies, but does not fall as low as the Zener voltage, then the current through the diode will vary but the voltage across the diode will be almost constant.

Varactor diodes

All diodes that use a junction between two layers of silicon have a measurable capacitance between anode and cathode terminals when the junction is reverse biased, and this capacitance

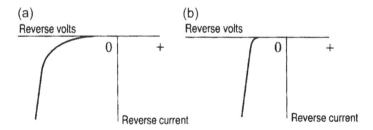

Figure 2.20:
Zener diode. The true Zener effect causes a 'soft' breakdown (a) at low voltages; the avalanche effect causes a sharper turnover (b)

varies with the size of the reverse voltage, being least when the reverse voltage is high (which could mean voltage levels of 6 V or less). This variation is made use of in **varactor** diodes, in which the doping is arranged so as to provide the maximum possible capacitance variation consistent with high resistance. A typical variation is of 10 pF at 10 V bias to 35 pF at 1 V reverse bias. Varactor diodes are used for electronic tuning applications because they behave like capacitors whose capacitance can be changed by altering the steady voltage bias across them.

Photodiodes, light-emitting diodes, and others

A **photodiode** is a form of diode that has its semiconductor junction (the area where the doping of the silicon changes) exposed in a transparent casing. The effect of light on a junction that is reverse biased is to allow some current to flow, so that the photodiode can be used to detect a light beam. One typical application is to detect the light signals in a compact disc (CD) player (see Chapter 12).

Light-emitting diodes (**LEDs**) use compound semiconductors such as gallium arsenide or indium phosphide. When forward current passes, light is emitted from the junction. The color of the light depends on the semiconductor material used for the diode and the brightness is approximately proportional to the size of forward current. LEDs with high output powers are now possible, and are used in low-energy lighting, particularly in modern cars.

■ Note

For more specialized applications, infrared and microwave diodes of various types can be obtained which emit radio or light waves when forward biased. The microwave types need to be enclosed in a suitable **resonant cavity**, which is a hollow metal box that acts like a resonant circuit for microwaves. Infrared-emitting diodes are widely used in remote controls.

■

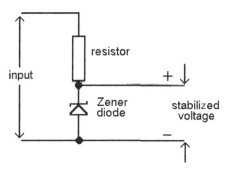

Figure 2.21:
The simplest application circuit for a Zener diode, making use of the diode to provide a stable voltage despite variations in input voltage or load current, provided that the current through the diode does not drop to a low value

Tunnel diodes are doped so as to have a characteristic which contains a portion with a reverse slope, indicating negative resistance in this region (so that the voltage across the diode rises when current increases). When a tunnel diode is biased into this unstable region it will generate oscillations at whatever frequency is determined by the components connected to the diode — RC, LC, or cavity — making this a versatile microwave oscillator for very low power outputs.

CHAPTER 3

Active Components and Integrated Circuits

Transistors

Modern active components are all based on **transistors**, which were invented in 1948 by Brittain, Bardeen, and Shockley. Before transistors came into extensive use, the main active components for electronic circuits were vacuum tubes (called **valves** in the UK), and these are still used for high-power transmitters. There are two main types of transistors, called **bipolar transistors** and **field-effect transistors** (**FETs**), and since the bipolar type came first into use, we will start with that. Most modern electronics systems now, however, are based on the FET, usually in integrated circuit (**IC**) form.

The transistor is a component that is created using a semiconductor crystal, and in a sense it is the inevitable result of the use of crystals in radio reception in the 1920s, because this started a line of research into crystal behavior which led to the transistor, even though the materials are quite different.

■ Note

It also had the much less desirable effect of starting a cult for imagining that crystals had psychic powers, showing that superstition and ignorance will latch on to anything to promote their nonsense.

■

Definition

A transistor is a semiconductor component with three terminals. An input between two of the terminals can alter the amount of current flowing to or from the third terminal.

This book is about electronics fundamentals, so that the way in which the transistor works is beyond our scope, but we are interested in what the transistor is and what it does. The symbol that is used on circuit diagrams for a bipolar transistor is a useful reminder (Figure 3.1). This shows three connections, labeled as **emitter**, **collector**, and **base**, with an arrowhead on the emitter lead that points in the (conventional plus to minus) direction of current.

Electronics Simplified. DOI: 10.1016/B978-0-08-097063-9.10003-2

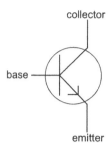

Figure 3.1:
The circuit symbol for a transistor, in this case the NPN type which will pass current when both collector and base are at a positive voltage compared to the emitter

For the type of transistor (known as NPN) that is illustrated in Figure 3.1, a steady voltage can be connected, with the positive end on the collector lead and the negative (earth) end on the emitter. No current passes, and the same would be true if the voltages were reversed, so that in this state the transistor behaves like an insulator.

With the collector positive and the emitter negative, if the base connection is now slowly made more positive than the emitter, a very small current will eventually pass between the base and the emitter. This pair of connections, emitter and base, is a diode. What is much more significant, however, is that when a small current passes between the base terminal and the emitter terminal, a much larger current will pass between the collector and the emitter. By much larger, I mean anything from 30 to 1000 (or more) times greater. The action is that a very small amount of current passing between base and emitter can control a much larger current passing between the collector and the emitter.

■ Note

The name transistor comes from **transfer resistor**, the original name for the device. The abbreviation **BJT** (bipolar junction transistor) is also used now, and is preferable because it distinguishes this type of transistor from the field-effect type that we shall look at later.

■

Suppose now that we use a resistor connected to a steady voltage to pass a very small steady current through the base. This will, as we have seen, make a much larger current flow through the collector to the emitter. The transistor is acting as a steady-current amplifier, making a large copy of a small current. If now we connect a signal waveform between the base and the emitter, this will cause the current in the base to fluctuate. The current in the collector circuit will also fluctuate, but with a much larger amplitude. For example, a current fluctuation of 1 microamp in the base circuit might cause a fluctuation of 1 milliamp in the collector circuit, a **current gain** of 1000 times.

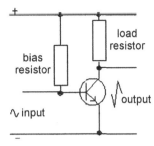

Figure 3.2:
The simplest (and least satisfactory) form of transistor amplifier circuit, showing input and output waveforms. The output wave is inverted compared to the input and is of greater amplitude

The transistor is now acting as a signal current **amplifier**. The steady current passing between base and emitter is called the base **bias** current. If we did not use this bias current, the base would not pass current until the input wave voltage reached about +0.56 V, so it could not have any amplifying effect on waves of small amplitude. With enough bias to allow current to flow, the transistor is a very effective amplifier for small signal inputs.

We can also make the transistor act as a **voltage** amplifier. The simplest (and least satisfactory) circuit is shown in Figure 3.2. A resistor is connected as a **load** between the collector and the positive supply terminal – nothing wrong with that because it is a normal way of converting a current signal into a voltage signal. A much larger value of resistor is connected between the base and the positive supply so that a small steady current flows through the base. Now if we add an alternating signal input at the base, taking care that the input is never so large that it causes the current to stop (by opposing the steady current), then the output will be a much larger voltage output.

This needs some thought. Imagine that the positive peak of the input wave has increased the base current. This will increase the collector current, and because there is more current through the load resistor the voltage across the resistor will be greater. Because there is more voltage across the resistor there is **less** between the collector and the emitter, so that the output wave is at its lowest, its negative peak. The output wave is a mirror image of the input, an inverted wave, as is indicated in the drawing.

The snag is that the bias current of this arrangement will not remain constant. A large value of resistance is needed (because the current needed at the base is very small) and even the smallest changes in the resistance or of the gain of the transistor (due to temperature changes) can make a large change to the bias. A much more practical circuit is illustrated in Figure 3.3, using two resistors to set a constant voltage at the base and another resistor in the emitter connection to stabilize the current.

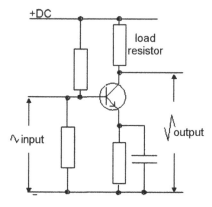

Figure 3.3:
The more usual form of a single transistor amplifier. The use of additional resistors stabilizes the steady currents that flow through the transistor

Summary

A transistor of the bipolar type (BJT) has three terminals: base, emitter, and collector. A small current passing between base and emitter will cause a large current to pass between collector and emitter, with both base and collector positive in this example (NPN transistor). This current gain effect can be used to construct a voltage amplifier in which the output is an amplified and inverted version of the input.

This principle of using a transistor and a load is at the heart of most of the electronic circuits (called **analog** or **linear** circuits) that we were familiar with before digital circuits appeared. Transistors are even better suited to digital circuits, as we shall see in Chapters 9 and 10. The snag about using transistors for amplification is that the output is never a perfect copy of the input (though the imperfections can be made to be very small). A graph of the signal voltage output against the signal voltage input is not a straight line so that this is not a perfectly linear amplifier. There are ways of improving this situation, which will be discussed in the following chapter. The ultimate solution is to use digital signals, and we will look at that later.

NPN and PNP Transistors

Bipolar transistors can come in two forms, called **NPN** and **PNP**. These P and N letterings indicate the type of carrier that takes most of the current in each region, with N meaning electrons as carriers and P meaning holes as carriers. The NPN transistor conducts mainly by movement of electrons in the collector and emitter and mainly by using holes in the base region. The practical effect is that the NPN transistor is used with a positive supply to both the base and the collector; the PNP transistor is used with a negative supply to both base and collector. Many circuits use a combination of PNP and NPN transistors.

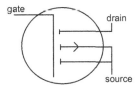

Figure 3.4:
The symbol for a metal-oxide-semiconductor field-effect transistor (MOSFET), in this case the type
called a p-channel MOSFET

Field-Effect Transistors

FETs have been available for almost as long a time as the bipolar type, but they were not
extensively used until later. Nowadays, the field-effect type is used to a much greater extent (in
ICs; see later) than the bipolar type, and the most important field-effect type is the metal-oxide-
semiconductor field-effect transistor, abbreviated to MOSFET.

The MOSFET uses quite different principles. Current can pass between two terminals called
the **source** and **drain**, and this current is controlled by the *voltage* (not current) on a third
terminal, the **gate**. On a circuit diagram, the MOSFET is indicated as shown in Figure 3.4, and
the important point about it is that there is no current passing to or from the gate for either
positive or negative bias.

This type of transistor needs a voltage signal only, not a current signal, so that the power needed
at the input is very much smaller than is needed for a bipolar transistor, almost a negligible
amount. Depending on the design of the MOSFET, a bias voltage can be used or the MOSFET
can be operated without bias. A typical amplifier circuit is illustrated in Figure 3.5.

MOSFETs do not provide as much amplification (voltage gain) as bipolar transistors, and their
uses were at one time predominantly for digital IC circuits (see later). Specialized types of
MOSFETs are now used in audio amplifiers and in tuning circuits for FM radios.

■ Note

There are various varieties of MOSFET, such as PMOS, NMOS, and CMOS, all of which
are used extensively in digital circuits. The differences are not important at this stage. The
MOSFET is sometimes known as IGFET, with IG meaning **insulated gate**.

■

Summary

The MOSFET is a form of field-effect transistor which has become the most commonly used type
of transistor. There are three terminals, called source, gate, and drain, with the voltage on the
gate controlling the current between the source and the drain. The current flowing in the gate is
almost immeasurably small. An amplifier circuit uses a load resistor connected to the drain, and
a voltage bias supply to the gate if needed.

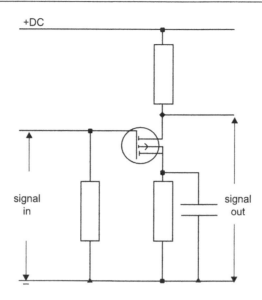

Figure 3.5:
A typical metal-oxide-semiconductor field-effect transistor (MOSFET) amplifier circuit

Switching

The analog uses of transistors, BJT or MOSFET, for amplifier circuits require the designer to set the correct bias, and use circuits that will provide the nearest approximation to a straight-line graph of output plotted against input. These amplifier circuits all cause the transistor to dissipate power, because the optimum bias is usually with the collector or drain voltage set to about half of the steady voltage supply, and passing a current that is about the amount that needs to be supplied. For example, if the DC supply is at 12 V and a current of 50 mA needs to be supplied, then the transistor will be biased to a collector or drain voltage of 6 V and a current of about 50 mA. This corresponds to a power dissipation of 6×50 mW $= 300$ mW. This is typically the maximum that a small transistor can dissipate without overheating.

All linear amplifier circuits suffer from this dissipation problem, and when higher power outputs are needed transistors have to be designed so that they can release their heat through a metal casing, or a metal stud, to metal fins that will dissipate the heat into the air. For a few specialized purposes, water cooling can be used, but air cooling, with or without a fan, is much more common.

There is, however, another way of operating transistors, though it is a long way removed from their use as linear amplifiers. If a transistor of any type is operated without bias, it will not pass current when there is no input signal. With no current flowing, there is no power dissipation, and so no heat to worry about. If a signal at the input (base or gate) now makes the transistor conduct fully, a comparatively large current will pass, but the voltage at the collector or drain will be low (as low as 0.2 V for some types) and the power dissipation will be small.

■ Note

If we use transistors for working with pulses, then the power dissipation will be very low, particularly if the transistor is switched on for only a short time in each cycle. This switching use of transistors is not suited to linear amplifiers, but it is ideal for digital circuits, as explained in Chapter 10.

■

Transistor switching circuits can also be devised that make the dissipation even lower. For example, we have assumed so far that the load for a switching transistor will be a resistor, so that power will be dissipated in the resistor whenever current flows. If the load resistor is replaced by another transistor in a switching circuit, the 'load' transistor can be switched as well, ensuring that its dissipation is also very low.

Summary

Transistors that are used for analog designs (the amplification of waves as distinct from pulses) are not as linear as we would like, and they dissipate power, risking damage to the transistors. The alternative way of using transistors is as switches, working with pulses that turn the transistor on for brief periods, so that linearity is not important and dissipation can be very low. This method of operation, however, is suited only to digital circuits.

Integrated Circuits

Only a few years after the invention of the bipolar transistor, G.R. Dummer, working at a UK government laboratory, suggested that the processes that were used to manufacture individual transistors could be used also to manufacture resistors, capacitors, and connections on the same piece of semiconductor material, making it possible to produce a complete circuit in one set of operations. Though the mandarins of the UK civil service who read his report attached no significance to it, others did. The USA was desperately trying to improve the reliability of electronic equipment for space use, and Dummer's suggestion was the answer to the problem.

It is possible to make individual electronic components reasonably reliable, but the weak point in large and elaborate circuits is the number of connections that have to be made between components. If you can replace 20 individual components by a single component then, other things being equal, you have made the circuit connections 20 times more reliable, and this is what creating a complete circuit in one set of operations amounts to. In addition, there is nothing that restricts the idea to just 20 components in a circuit, though the pioneers could not have imagined creating circuits with several million components on one small piece (**chip**) of silicon. This device is the **integrated circuit** or IC. As is the case with most British ideas, we ended up buying into the technology that we had failed to develop for ourselves.

The space race between the USA and the USSR spurred the development of ICs, so that by the 1970s we could buy complete circuits on a single tiny chip of silicon that were vastly more reliable than anything we had ever imagined. These ICs were mainly for digital circuits, though there are also many types of linear (amplifier) ICs. In addition to reliability the advantages of these ICs include low cost, small size, low dissipation, and predictable performance. By this point in the twenty-first century, making connections between individual (**discrete**) transistors is an almost forgotten skill used only where circuits are created for special purposes.

Summary

The IC was originally a British idea, but you could be pardoned for not knowing that because they made nothing of it. The principle is to improve reliability by using the manufacturing methods that are used to create transistors also to make resistors, capacitors, and connections between these components, all on the same tiny chip of silicon that is used to make transistors. These chips are mass produced, and the elimination of external connections makes the reliability of a circuit with a large number of individual components as good as the reliability of a single component. In addition, the cost can be low because of mass production, and the size of a complete circuit is little more than that of a single transistor.

Linear Integrated Circuits

Digital circuits have always accounted for a majority of the ICs that have been manufactured, but there has always been one particularly important type of linear IC, called the operational amplifier (or **opamp**). The circuit of an opamp is unimportant to the user, and only the designer is likely to know exactly what goes on inside the chip, so the user of an opamp works from a set of figures that describe its performance. The usual symbol is illustrated in Figure 3.6.

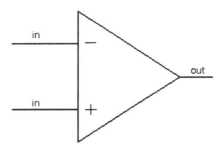

Figure 3.6:
The symbol for an operational amplifier. The + and − input markings refer to the phase of input signals, not to the voltage of DC supplies. The DC supplies are usually balanced, typically +12 V and −12 V

Instead of spending hours on the design of an amplifying circuit, a designer can use an opamp and (typically) two resistors to get the amount of gain that is required, and the calculations are simple. This does not mean that opamps can be used for any task that was formerly carried out by circuits using separate transistors (discrete circuits), but 99% of them would be a reasonable estimate, sufficient to make the opamp a considerable boon.

■ Note

This is an excellent example of how passive components (resistors in this case) can be used to control the action of an active component (the opamp). Because the amount of gain is set by the passive components, their tolerance and stability are very important.

■

The typical opamp has two inputs, marked + and −, and one output, and it is used with a balanced pair of DC supplies, typically +12 V and −12 V. The input markings do not refer to + or − supplies, but to the **phase** of signals. If you use the + input, the output signal will be in phase with the input signal. If you use the − input, the output signal will be in anti-phase, a mirror-image waveform, like the output from a single-transistor amplifier. For a number of reasons, it is much more usual to take the input signal to the anti-phase (−) input.

Figure 3.7 shows a typical amplifier circuit that will provide a gain of about 47 times. How do we know? Because that is the ratio of values R_1/R_2. The opamp circuit is unknown, but the manufacturer guarantees that the gain of the chip is at least 100,000. The resistor R_1 is connected between the output and the input, so that some of the output is fed back, opposing

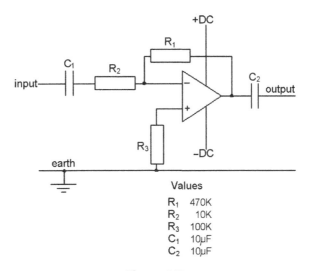

Figure 3.7:
A typical opamp circuit in which the gain (amount of amplification) is determined by the values of resistors R_1 and R_2

the input. This reduces the gain, and it also makes the amount of gain depend only on the resistor values, not on anything inside the IC (unless some part of the IC circuit is overloaded). This means an end to complicated calculations and guesswork, and even the arithmetic is simple.

■ Note

In this example, the DC power supplies have been shown. Opamp circuits are usually shown with all of these power supply connections omitted. The resistor R_3 is used to ensure that the + input is connected to earth, and its precise value is not important.

■

Using an opamp is not an answer to everything, because the circuit might need to work at a frequency higher than the opamp can cope with, but a later type of opamp probably will cope. In addition, there is a host of specialized opamps designed for specific purposes, and catering for almost all the exceptions to the general rules.

There is one serious handicap that affects all opamps. Because an opamp is designed for amplification, the transistors must use bias and therefore dissipate heat. This determines how complicated an opamp can be, in terms of the number of transistors in the circuit, because each transistor will contribute to the heat output and that heat must be transferred to the air if the opamp is not to overheat. The same applies to opamps that are intended to provide a power output to loudspeakers, electric motors, and other devices.

Summary

The most common type of linear (analog) IC is the operational amplifier or opamp. This has a very high gain value, and is normally used in a circuit in which two external resistors control the gain. In the situations where such an opamp is not suitable, there are other designs that will deliver the performance that is needed for more specialized purposes. The snag is that heat dissipation limits the complexity and power capabilities of an opamp.

Digital Integrated Circuits

Digital ICs are the more common variety, mainly because of the vast number of digital devices (not just computers) that make use of these types of ICs. The transistors inside digital ICs are being used not as amplifiers, but as switches. This means that the heat dissipation for each transistor is very low, allowing digital ICs to be constructed using hundreds, thousands, and even millions of transistors. In addition, heat-dissipating components (resistors) can be designed out because substituting a transistor for a resistor is easy when both use the same techniques (and an IC transistor can be physically smaller than a resistor). Passive components are much less important in digital circuits than in analog circuits.

Definition

Digital ICs deal with pulse inputs and outputs and use switching actions with very low dissipation.

The simplest digital ICs carry out just one type of switching action, and they can perform the operations that are called **logic actions** (Chapter 10). The type of circuits that can be constructed using these ICs would typically be used as controllers for machines, using several inputs to decide whether an output should be on or off.

When should your washing machine start a cycle? Obviously, it is when the main switch is on, a program is selected, the drum contains some clothes, the water supply is switched on, and the main door is closed. The machine must not switch on unless all of these 'inputs' are present, and this action of providing an output only for some particular set of inputs is typical of the type of circuits we call **combinational**. We will come back to all that in Chapter 10. All of the first generation of digital ICs were intended to solve that type of problem, and these ICs are still in production more than 40 years on.

The next development was to make ICs that dealt with **sequential** actions, such as counting. These ICs required more transistors in each circuit, and as manufacturing methods improved designers found that they could produce not just the ICs that could be assembled into counters but the complete counters in IC form. At the same time, IC methods were being used to create displays, the LED and LCD displays that are so familiar now, so that all the components that were needed for a pocket calculator were being evolved together, and soon enough a complete calculator could be made using just one chip.

The pocket calculator story is a useful one to trace this part of the history of electronics. The first pocket calculators used several ICs, and they needed a considerable amount of assembly work. You could at that time buy DIY kits if you were curious to find out how the calculator was assembled, and such kits were also cheaper than a complete calculator. Nowadays, the calculator consists of just one IC, and there is practically no assembly. It costs more to assemble and package the components as a kit than to make and package the complete calculator, and costs are so low that the calculators can often be given away as a sales promotion.

Another thread of the story concerns the power required. The first pocket calculators needed four AA cells and gave about one month of use before these were exhausted. Power requirements have been so much reduced that some calculators are likely to be thrown away before the single cell that they use is exhausted, and it is possible to run calculators on the feeble power from a photocell (which converts light energy into electrical energy).

The early digital ICs were constructed using bipolar transistors, mainly because at the time these were easier to construct. The snag with bipolar transistors is that they need current inputs: no current flows between collector and emitter unless a current flows between base and emitter. The base current might be small, but some base current must exist, and so a bipolar transistor must inevitably dissipate more power than the MOSFET type, which needs no current between gate and source terminals.

Eventually, then, digital ICs started to be manufactured using MOSFET methods, and this allowed the number of transistors per IC to be dramatically increased. This packing of transistors was measured roughly by names that we use for **scale of integration**. This is described in terms of the number of simple logic circuits (**gates**) that can be packed into a chip, and the first ICs were small-scale integration (SSI) devices, meaning that they contained the equivalent of 3–30 logic circuits. The pace of development at that time (the 1960s) was very fast, so that the terms medium-scale integration (MSI) and large-scale integration (LSI) had to be introduced, corresponding to the ranges 30–300 and 300–3000 logic circuits, respectively.

This is a good example of how technology runs ahead of expectation. LSI soon became commonplace, and we had to start using very-large-scale integration (VLSI) for chips with more than 3000 gates per chip. Soon enough, chips containing 20,000 or more gates were being manufactured, but a new label, extra-large-scale integration (ELSI), was not introduced until more than one million gates could be put on a single chip. Scale-of-integration names are not normally used nowadays. Moore's law once predicted that the number of transistors that could be placed on an IC would double each year, dating from the start in 1958. Gordon Moore (founder of Intel) thought in 1965 that this trend would flatten after 10 years, but it holds true at the time of writing in 2010 and may continue as new ways of producing ICs are developed.

Summary

Digital ICs are classed in terms of the number of simple gate circuits that they replace on average. Modern chips are usually of the VLSI class, equivalent to 20,000 or more gates, and computer ICs are often in the ELSI class, equivalent to one million or more gate circuits.

The Microprocessor

The development of the microprocessor followed logically from the use of digital logic circuits, and this is an introduction only; more details follow in Chapter 13.

The microprocessor is a kind of universal logic circuit. Imagine you have a large number of digital logic ICs. You could make these into circuits, each of which would carry out some sort of control action, depending on how you connected the chips together. You might find that the circuits you used had a lot in common, and that you could make one circuit which used a number of switches, so that by setting switches you could change the overall action from one design to another. The next logical step is to make these switches in the form of transistors, so that the switching can be electrical. It is that this stage that the microprocessor becomes a possibility.

The first microprocessor was designed and built to a US military contract which called for a logic circuit whose action could be controlled by inputs of signals called **programming signals**. The military contract was canceled, and the manufacturer (Intel) was left with a large number of devices for which there was no known market. A few enthusiasts found that these chips could be used to make simple computers, and once this became known, a small firm called Altair put together a kit for building these computers. The sales of this kit were phenomenal, and by the end of a year most of the microcomputer firms that we know today were in business. It is ironic to note that larger firms such as IBM dismissed these machines as toys and predicted that they would all disappear inside a year. As it is, the IBM PC business was sold to Lenovo (based in China) in 2004.

Summary

The microprocessor is a form of universal logic circuit whose action can be set by the user. This allows the microprocessor to replace a vast number of other circuits, and to be more flexible in use, because its action can be modified by programming it, using pulse inputs.

CHAPTER 4

Linear Circuits

Linearity

We have seen that there are two main types of electronic circuits, and we label them as **analog** or as **digital** circuits. Though we can design and construct both types of circuits using the same set of active and passive components, the active components are used in very different ways and the waveforms that are processed are very different.

■ **Note**

Analog circuits are not necessarily linear — a rectifier circuit is just one example — and such non-linear circuits are not digital. We class these circuits as **non-linear analog circuits**. In short, all linear circuits are analog, but not all analog circuits are linear.

■

A linear circuit is a type of analog circuit that is designed to make a **scaled** copy of a waveform, and by scaled we mean that the amplitude of the output of the linear circuit is a fraction or a multiple of the amplitude of the input waveform. The output amplitude might be smaller, in which case we often call the circuit an **attenuator**. The output amplitude might be equal, in which case we usually call the circuit a **buffer**. Quite commonly the output amplitude is greater than the input amplitude, and the circuit is an **amplifier**. If the circuit is truly linear, the output waveform has the same frequency and the same **waveshape** as the input waveform — it is a true copy at a different amplitude scale (Figure 4.1), and the ratio of the output amplitude to the input amplitude is called the **gain**.

The name of **linear** circuit arises from the shape of graphs of output amplitude plotted against input amplitude. For a perfectly linear amplifier, this graph should be a straight line, hence the name *linear*. As it is, because of the imperfections of active components, amplifiers are never perfectly linear, though we can obtain very good linearity in a buffer circuit, and perfect linearity in an attenuator which uses only passive components.
Figure 4.2 shows perfect linearity and some common varieties of imperfections in graphs for amplifiers.

Electronics Simplified. DOI: 10.1016/B978-0-08-097063-9.10004-4

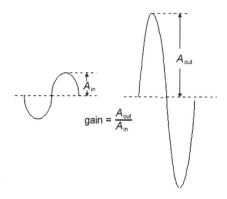

Figure 4.1:

Amplitude and gain. The amplitudes must be measured in the same way, and peak amplitude measurements are illustrated here

■ **Note**

A graph cannot show small amounts of non-linearity, and we have to use instruments to detect traces of non-linearity by other methods. One such method is to use a perfect sinewave as an input; any non-linearity will cause the output to contain harmonics, waves at higher frequencies, and these can be detected by sensitive instruments.

■

Definition

A linear circuit is one for which a graph of output plotted against input is a straight line. Linear circuits are used in analog designs, though not all analog circuits need be perfectly linear.

The most common imperfection is curvature: the graph line is curved rather than straight. This means that the amount of gain, the scaling, alters as the input amplitude alters. For example, large-amplitude signals may be amplified less than small-amplitude signals. This type of non-linearity is sometimes required, and the human ear itself is non-linear, a way of protecting us from the worst effects of the noise of explosions, aircraft noise, and discos.

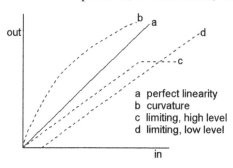

Figure 4.2:

Linear and non-linear behavior of amplifiers

In general, though, for purposes such as sound reproduction, we need amplifiers that are as linear as we can get them, and a target of 0.1% deviation from a straight-line graph is taken as a reasonable target for a hi-fi amplifier. This deviation from linearity is often called the **distortion figure**, and quite high distortion figures (of 10% or more) are common for radios, compact disc (CD) players, guitar amplifiers, disco amplifiers, and other equipment for which high-quality reproduction is not an important factor.

Another imperfection that is illustrated in the drawing is **limiting**. When an amplifier limits, the output amplitude stays constant even though the input amplitude is changing. This causes the shape of the waveform at the output to be severely distorted, flattening at the tops as in Figure 4.3. All amplifiers will limit if the input amplitude is too large, so that a complete amplifier system has to be designed so that limiting cannot occur even if the volume control is turned full up. If an amplifier is designed to have different input amplitudes, as it must if it is supplied from different sources such as tape-heads, pickups, radio circuits, CD players, etc., then there should be adjustments provided so that the manufacturer or user can set each input to the same level so that the master volume control does not need to be adjusted when you switch from one source to another. It is unusual to find these adjustments provided except on expensive sound equipment.

The third type of non-linear behavior illustrated is another form of limiting. The amplifier simply does not amplify signals that are at a very low-amplitude level. This type of non-linearity has in the past been used deliberately to reduce the noise signals from cassette tapes but, like the other types, it is undesirable in normal use. When transistor amplifiers first appeared, a form of this type of distortion, called **cross-over distortion**, was a common fault which delayed the acceptance of transistors for high-quality linear amplifiers until solutions were found.

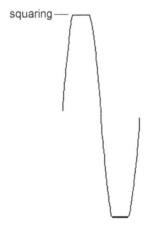

squaring —

Figure 4.3:
Squaring is the effect on the waveform of some type of limiting action in an amplifier

Summary

Non-linear behavior includes curvature and limiting. Though these effects are sometimes deliberately used, high-quality amplifier designers aim to reduce non-linearity to as low as can economically be attained. The amount of non-linearity is often expressed as the percentage by which the graph line deviates from a straight line, and a usual target for high-quality amplifiers is 0.1%, one part in 1000 (though this figure cannot be obtained from a graph).

■ Note

Some devotees of hi-fi are not necessarily logical in their choice, and some prefer to hear the type of distortion (often large) that is caused when the old-fashioned vacuum tube amplifiers are used. Others are prepared to believe that they can hear differences caused by the type of wire used for connecting cables. One more skeptical authority has commented that the only significant difference cables can make to loudspeaker performance is if they are too short to reach the loudspeakers. The most sensible advice is to avoid the type of magazines that use the sort of fancy descriptions that are employed by wine tasters or art reviewers.

An attenuator that is constructed entirely from passive components, such as the type shown in Figure 4.4, is perfectly linear for the normal signals that we are concerned with. This is because non-linearity is caused almost exclusively by active components. Passive components contribute to non-linearity only if they are overloaded, and that is reasonably easy to avoid.

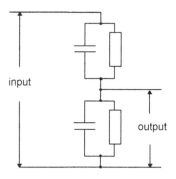

Figure 4.4:
An attenuator formed from passive components does not cause any non-linear distortions. The type illustrated here is a compensated attenuator which will work over a very wide range of frequencies

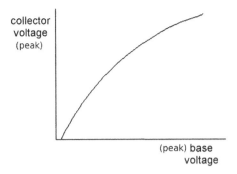

collector
voltage
(peak)

(peak) **base**
voltage

Figure 4.5:
The non-linear behavior of a transistor at its worst. Correct bias and choice of transistor type can help, but the graph shape is always a curve. Note that what is being plotted is the peak amplitude of the waves at input and output

The root causes of non-linearity in bipolar transistor circuits are:

*   A simple plot of collector wave voltage against base wave voltage for a transistor in a simple amplifier circuit (Figure 4.5) is curved.
*   There is no collector output for small base waveforms.

The same considerations apply to metal-oxide-semiconductor field-effect transistor (MOSFET) circuits, though MOSFETs can be manufactured so that they have more linear input/output graphs. Careful attention to transistor biasing and circuit design can result in a graph shape that is closer to a straight line, eliminating the region where there is no collector or drain output, but there is always some curvature. The problem is tackled by cunning circuit design that applies corrections to the output wave (a system called **negative feedback**), but this is not in itself a way of making a poorly designed amplifier perfect; though it can make a good design work better.

■ Note

Good design of operational amplifiers can also reduce distortion to a very low figure provided that there is no overloading in any part of the amplifier.

■

Yet another type of distortion is called **slew-rate limiting**, and it is quite different from the others. Slew-rate limiting occurs in a transistor, whether it is a separate transistor or part of an integrated circuit (IC), and it cannot be eliminated by circuit design, only by using transistors that have better design characteristics. Imagine a simple single-transistor amplifier. For a small input this may be able to cope with a waveform that has a sudden change in voltage, but if the change in voltage at the input is large the transistor simply cannot pass enough current at the output to charge or discharge stray capacitance. The effect is that the change in voltage at the output takes longer or is limited. This effect is most likely to be a problem for power amplifiers, and the solution is either to filter out any inputs that have large sudden voltage changes, or to use different types of semiconductors in the active part of the circuit.

Summary

Slew-rate limiting affects amplifiers that handle inputs with sudden voltage changes where the change of voltage can be large. The effect causes a form of distortion (because the waveshape is changed) that can be eliminated only by using better semiconductor types.

A **buffer** circuit makes use of an active component circuit controlled by passive components, but with the output amplitude the same as (or slightly less than) the input amplifier. For this type of action, it is possible to make transistors work in a fairly linear way so that distortion can be very small, though never zero. The purpose of a buffer is to prevent **loading** of a signal. For example, suppose that the input to an amplifier is a signal of 1 mV which can supply only 1 μA of signal current. Now if we connect this to an amplifier which needs to pass 10 μA at its input, we are asking too much of the signal source, loading it, and this is certain to cause non-linear behavior.

If we place a buffer stage between the amplifier and the source, using a buffer which takes a current much less than 1 μA of current but can provide more than 10 μA at its output, then we avoid the loading effect and improve the performance of the whole amplifier. Buffers are used in all types of circuits, linear and digital, for this same purpose, to avoid taking more current from a signal source than it can comfortably supply. Another function of a buffer is to isolate two stages so that the signals in the second stage cannot affect the first stage.

The simplest type of transistor buffer circuit is illustrated in Figure 4.6, using a bipolar transistor. This is called an **emitter-follower**, and the output signal voltage is in phase with the input. The amplitude of the output is slightly less than the amplitude of the input, but the power output is more because the emitter current of the transistor can be much greater than the base current. This is an example of 100% negative feedback (see later in this chapter), because the input to the transistor is the voltage between base and emitter, but this is equal to the input

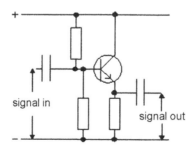

Figure 4.6:
An emitter-follower circuit, a simple example of a buffer that provides power gain without voltage gain

signal (between base and earth) minus the output signal between emitter and earth. The MOSFET equivalent is the **source-follower**.

Summary

Buffers are linear circuits with zero gain, and it is fairly easy to ensure that they are almost perfectly linear. Buffers prevent too much current being taken from the source of a signal, and are used to isolate one section of a circuit from the next. The emitter-follower type of circuit is a simple type of buffer, and an opamp equivalent, the voltage-follower, is widely used.

Gain

An amplifier carries out the action of making an enlarged copy of the waveform that is used as its input signal, and the ratio of the output signal to the input signal is called the **gain** of the amplifier. This figure of gain could be written, for example, as $\times 10$, $\times 100$, or even $\times 1000$, but we seldom use this way of expressing gain.

Definition

Voltage gain is defined as (output signal voltage)/(input signal voltage). We can also define **power gain** as (output power)/(input power), or **current gain** as (output current)/(input current).

Expressing gain as a simple ratio of voltages is often useful, particularly if we are aiming for some definite signal output level such as 0.5 V RMS (root mean square; see later), but for amplifiers that deal with signals in sound or vision systems, the **decibel** scale is more useful. There is nothing mysterious about it, except that it is often misused. We have known for more than a century that human ears and eyes do not have a linear response. For example, when the amplitude of two sound waves is compared, one with twice the amplitude of the other, your ears do not hear one sound as twice as loudly as the other. Similarly, two lights, one with twice the amplitude of the other, do not appear to your eyes so that one looks twice as bright as the other.

The type of scale that your senses use is **logarithmic**, meaning that the quantity we perceive is related to the logarithm of the wave amplitude. These logarithm units of comparison are called **decibels**, named after Alexander Graham Bell, who we remember for the invention of the telephone, but whose main interest in life was the problem of deafness; in fact, the first telephone was invented as an aid for the deaf.

Definition

If one voltage amplitude is 100 times the size of another, our senses tell us that when signal amplitude is converted into sound or light the ratio is something more like 40 times, which is 20 times the logarithm of 100. As a formula this is $20 \log (G_V)$, where G_V is the simple voltage gain. If this is applied to power gains, it becomes $10 \log (G_P)$, where G_P is the simple power gain.

A voltage amplifier is very likely to be specified with its figure of gain expressed in decibels, abbreviation dB. A decibel gain of 40 dB corresponds to a voltage gain of 100 (and a power gain of 10,000). Table 4.1 shows some values of voltage gain ratios and decibel amounts.

■ Note

Strictly speaking, decibels should be used only for comparing power levels, using the 10 log (G_P) formula. The use of decibels for comparing voltages, however, is so common that it cannot be ignored, though it is really valid only when the voltages are measured across the same value of resistance (which implies that the current has been amplified as much as the voltage).

■

The trap to watch for is that a decibel value always refers to a **ratio**. You cannot, for example, say that a noise is at 100 dB unless you specify what your comparison is. A figure of 100 dB means that the noise is that much louder than something else, but you have to specify what that something else is. There are agreed comparisons for sound, but they are based on measured power rather than something subjective like a demented fly at 20 feet.

The gain of an amplifier is a ratio and is therefore always specified in decibels, and this form is also used for attenuators. For example, you might specify a -10 dB attenuation, meaning that the output voltage is only about 0.3 times the input voltage. The minus sign indicates that this is a ratio of input to output, not output to input, with the output amplitude less than the input amplitude.

Table 4.1: Voltage gain ratios and decibel amounts.

Gain	dB
2	6
3	9.5
5	14
10	20
20	26
30	30
50	34
100	40
200	46
300	50
500	54
1,000	60
2,000	66
3,000	70
5,000	74
10,000	80

Note: Values have been rounded. You can use this table to find intermediate values by remembering that **multiplication** of gains equals **addition** of decibels. For example, a gain of 15 times is 5 × 3, which in dB is 14 + 9.5 = 23.5.

Summary

Because our senses work on a logarithmic scale, it is often more convenient to express gain in this way, and this is the purpose of the decibel scale. For power gain, the number of decibels is equal to 10 log (G_P), where G_P is the power gain, and this is the true definition of the number of decibels. For many purposes, however, it is convenient to work in voltage terms, and the number of decibels corresponding to a voltage gain is 20 log (G_V), where G_V is the voltage gain.

Frequency Response

The graph of output amplitude plotted against input amplitude shows up gross non-linear behavior, but there is another form of distortion that can affect any linear amplifier, and even passive devices such as attenuators. This is **frequency distortion**. Some amplifiers are intended to work with a single frequency or a narrow band of frequencies. These are tuned amplifiers and we shall look at them later in this chapter. More usually, an amplifier has to work for a range of frequencies, and it must have the same amount of gain for all the frequencies in that range, which is its **bandwidth**.

For example, an amplifier that is used along with a CD player should be able to deal with the frequency range of about 30 Hz to 20 kHz, the **audio range**. These frequencies represent the limits of a (young) human ear, and as we get older our ability to hear the higher frequencies is steadily reduced; one estimate is that the rate is 1 Hz less each day, so that the upper limit can eventually be as low as 10 kHz or less. The human ear is at its most sensitive at a frequency of around 330 Hz (the average frequency of the female voice), and sensitivity is reduced at the very low as well as the very high ends of the frequency range. An amplifier designed to deal with the normal audio frequency range is an **audio amplifier**. This is the most common form of amplifier because every radio, mobile phone, cassette, MP3 or CD player, and television receiver, includes an audio amplifier stage.

A typical graph of gain plotted against frequency for an audio amplifier is illustrated in Figure 4.7. The gain is displayed in dB units, and the frequency is shown on a logarithmic

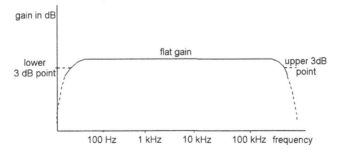

Figure 4.7:
A frequency response graph, showing the 3 dB points, between which the bandwidth is measured

scale, so that the markings are 1, 10, 100, 1000, and so on, at equal intervals. This type of graph is called a **frequency response** graph.

Definition

A logarithmic scale is one that uses a fixed *distance* to represent a fixed *multiple*. For example, you might use 1 cm on the graph to represent each ten-fold change in a quantity leading to a 1, 10, 100, 1000 scale, with these numbers at equal intervals. On such a scale, the mark midway between 10 and 100 does **not** represent 50 (it is closer to 31).

■ Note

If we used a linear scale, with the distance proportional to the frequency, the length of the frequency scale would be ridiculous. For example, if we allowed 1 cm for 10 Hz, the length of the scale in this example would be about 2000 cm. In addition, because the decibel scale is logarithmic, it makes sense to use this type of scale also for frequency. ■

In the example, the frequency response is level for most of the range, with a downturn at each end. The convention is to take the frequency response as extending between the points where the level is 3 dB below the flat (maximum) portion. The basis of this is that 3 dB is an amount of change that in terms of sound or light is only just significant and noticeable to ears or eyes, so we can ignore variations of less than this amount. Even if the frequency response graph has no flat portion, we can ignore variations that are less than 3 dB. One of the advantages of using the decibel scale is that this amount of variation looks small on the graph: a 3 dB variation in terms of gain means a two-fold change (twice or half), which would look very large on a linear graph, but does not have so much of an impact on ear or eye.

Definition

The bandwidth of an amplifier is defined as the range of frequencies between which the change in gain is 3 dB. For audio amplifiers, this is usually quoted in terms of a range of frequencies, such as 20 Hz to 20 kHz. For radio frequency amplifiers (see later) this will be expressed as the **difference** between the frequencies, such as 5 kHz, 100 kHz, 5.5 MHz, and so on.

No amplifier can have a perfectly flat frequency response over a really large range of frequencies. The use of capacitors to carry signals between sections of a circuit limits the gain figure for the low frequencies, because the reactance of a capacitor is high for a signal at a low frequency. There are also limits on the gain that can be achieved at high frequencies, caused by transistors themselves and by stray capacitance. Stray capacitance is unplanned capacitance between different parts of the circuit, and this arises because a capacitor is an insulator sandwiched between conductors, so that any two pieces of metal separated by air must have some capacitance. These strays filter off the highest frequencies because even a small value of capacitance will have a low reactance for signals at high frequencies. In some circuits the design of the circuit is able to compensate for such losses. In other circuits there may be hills and dales on the frequency response graph because of **resonance** effects caused by stray capacitances and stray inductances.

Broadband Amplifiers

There is a special class of linear amplifiers called **broadband** amplifiers. This name is reserved for amplifiers that have a much wider frequency response than the ordinary audio amplifier. For example, the waveform that carries the picture information (for an analog television receiver, for example) is called the **video** signal. Ideally, an amplifier for this signal should have a bandwidth of zero to around 5.5 MHz (**zero** means that changes in the steady voltage level are also amplified). In such a signal, there is a DC portion which carries the information on the overall brightness of the screen, while the highest frequencies carry the information on the finest detail.

This bandwidth is much greater than is used for audio signals, and to put it in perspective, it is more than five times as much as the whole of the medium-wave band on a radio. Even this, however, is small compared with some other bandwidths. A good computer monitor, for example, will use an 80 MHz bandwidth, which is why we once used (costly) cathode-ray tube (CRT) monitors for computers rather than (cheap) CRT television receivers, and why computer images that look so clear on a good monitor look so fuzzy on a television screen. Now that both monitors and television screens use flat-screen digital display methods the differences are much less. Some measuring instruments that are used to find the amplitude and frequency of waveforms need amplifiers with a bandwidth of 100 MHz or more.

These broadband amplifiers need to be designed with great care, and the physical arrangement (**layout**) of the components is just as important as the theoretical design of the circuit. Very large bandwidths are also used in connection with the amplifiers that are used for receiving radio signals, particularly the low-noise box (**LNB**) amplifiers that are contained in satellite receiving dishes.

Tuned Amplifiers

The broadband amplifiers that are used, for example, in a computer monitor or an oscilloscope are classed as **untuned** (or **aperiodic**) amplifiers, but we also make considerable use of **tuned** amplifiers. Recalling Chapter 2, a combination of an inductor and a capacitor is a tuned circuit that has a peak response at some frequency. This value of frequency, the tuned frequency, can be calculated from the amounts of inductance and capacitance, but the response of the circuit shows some gain for a range of frequencies around the tuned frequency. As usual, the

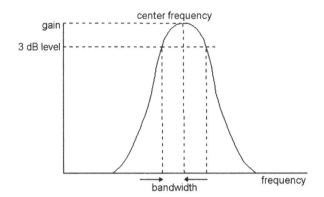

Figure 4.8:
The bandwidth for a tuned circuit is measured to one of the 3 dB points on either side of the tuned (center) frequency

bandwidth is calculated as the frequency range between the points where the response is 3 dB down (the −3 dB points) (Figure 4.8).

Definition

A tuned amplifier is one whose bandwidth is centered around a single frequency (or center **frequency**) at which the response is maximum. This is usually achieved using resonant circuits in conjunction with active components.

Bandwidth is important, because it affects the use that we can make of radio waves. As we will see in Chapter 6, a radio transmission makes use of a small band of frequencies that lie around a central maximum which is the frequency set at the transmitter (the **carrier frequency**). This bandwidth for medium-wave broadcasts is only about 5 kHz on each side of the tuned frequency. This, in turn, means that the frequency range for sound on medium-wave radio is only about 5 kHz. This situation has arisen because too many transmitters use the medium waves. As far as speech is concerned, this 5 kHz is adequate, but it is nothing like adequate for the good reproduction of music (or the reproduction of good music). In addition, the older methods of using a radio wave to carry sound signals (modulation methods) are very susceptible to interference both from natural (e.g. lightning) and artificial (e.g. car ignition systems) causes.

This problem has been around for a long time, and it was first solved by an amazing US genius called Armstrong (whose other inventions have also been landmarks in radio). In the 1930s, Armstrong developed a method, called **frequency modulation (FM)**, for carrying high-quality sound on radio waves. This, however, requires a bandwidth of about 100 kHz, so that we need to be able to construct tuned amplifiers with this range of bandwidth for FM radio use. The size of the bandwidth also makes FM unsuitable for medium-wave broadcasts, and carrier frequencies in the 90–110 MHz range have been used.

The most recent solution for high-quality radio broadcasting is **digital radio**, and this makes much more efficient use of bandwidth than FM (estimated at about three times more efficient). Digital radio broadcasting allows a bandwidth to carry more than one broadcast so that a set of different channels can be accommodated. Television requires considerably larger bandwidths, and a typical (analog) television broadcast might use a tuned frequency of 800 MHz, using a bandwidth of about 6 MHz. Digital television can, contrary to what you might expect, use smaller bandwidths because of the use of **compression** and multiplexing (see Chapter 16).

The simple tuned circuit, as noted in Chapter 2, cannot easily provide such large bandwidths, so various circuit tricks have been used to broaden the bandwidth. All of these methods sacrifice gain so as to obtain more bandwidth, and the simplest is to connect resistors across tuned circuits. These are called **damping** resistors, and their action, as illustrated in the graph of Figure 4.9, is to reduce the gain of the tuned circuit but broaden the bandwidth. The reduction in gain can, if necessary, be overcome by increasing the number of tuned amplifying stages.

Another method of increasing bandwidth but maintaining gain is to use several tuned stages, but with each tuned to a different frequency around the central frequency. This is called **stagger tuning** and, along with damping, it can provide the wide band amplification that is needed for the reception of television signals (the broadband amplifier for video is used at a later stage; see Chapter 8).

One important point about tuned amplifiers is that they need not be particularly linear, though they are classed as linear amplifiers. Any tuned amplifier uses a tuned circuit as a load, and even if the amplifier limits, cutting off part of the wave, the tuned circuit will complete the wave. This is because a tuned circuit acts like a pendulum which, once set into motion, keeps swinging. The tuned circuits are used as loads, and once a transistor has started to pass a wave

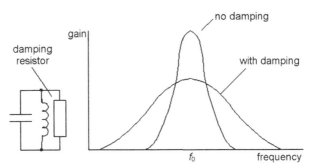

Figure 4.9:
The effect of adding a damping resistor to a tuned circuit is to broaden the bandwidth and reduce the gain

of current through the tuned circuit, the voltage across the tuned circuit will be a complete wave even if the transistor stops passing current for part of the time.

Summary

Tuned amplifiers use resonant circuits tuned to a particular frequency, and the circuits can be arranged to give whatever bandwidth is required. Quite simple circuits can provide narrow bandwidths, but for wide bandwidths or several megahertz more elaborate designs are needed.

Power and Root Mean Square Values

Using a linear voltage amplifier on a feeble signal will result in an output that is a signal at a much higher voltage level, but we cannot necessarily use this signal to operate devices such as a loudspeaker or an electric motor. The reason is that these devices need a substantial amount of current passed through them as well as the voltage across their terminals, and a voltage amplifier cannot supply large currents. What we need is a **power amplifier**.

For example, a voltage amplifier might provide an output that was a voltage wave of 6 V which could supply no more than 1 mA of current. The power amplifier might provide a wave with a voltage of 6 V and a current of 2 A.

■ **Note**

The name of **power amplifier** is misleading, because **any** amplification of voltage (without reducing current) or current (without reducing voltage) is power amplification. In fact, the power amplifier is usually a current amplifier, but since it is used to supply power to devices like loudspeakers, the name of power amplifier is more common.

■

Power amplifiers are used also in applications that have no connection with loudspeakers. For example, the old dot-matrix printer for computers uses a set of power amplifiers to drive the pins in the print-head. There are usually nine or 15 such pins, and each is moved by passing a current through a coil that surrounds the pin. This action requires power (the print-head dissipates a considerable amount of heat and can become very hot), so that a power amplifier is used to supply each coil. Modern ink-jet printers also have to have their heads driven by power amplifiers, because the load is either a tiny heating coil or a piezoelectric transducer (producing a squeezing action when an input pulse is applied). Though ink-jet and laser printers are dominant in home computing, dot-matrix types are still used in miniature printers and in some office work because carbon-paper copies can be created from a dot-matrix printer.

The voltage gain of a power amplifier is often very low, often less than unity, so that the output voltage is less than the input. The current waveform, however, has a much greater amplitude at the output, and current gains of 1000 or more are common. Because voltage gain is unimportant, it is easier to make a power amplifier quite linear for small signal amplitudes, but quite another matter to make it linear over the whole range of signals that it must cater for, particularly when the output is connected to something such as a loudspeaker which behaves like a complicated circuit of resistors, capacitors, and inductors.

Summary

Power amplifiers are usually current amplifiers that are needed when a load, such as a loudspeaker or electric motor, has to be supplied with a waveform that has been electronically generated.

Calculating Power

The calculation of power for waveforms is another matter we need to look at. For steady voltages and currents, the power dissipated in a resistor is easily calculated by multiplying the value of current through the resistor by the value of voltage across the resistor. In symbols, this is $V \times I$, and when units of volts and amps are used for voltage and current, respectively, the power is in units called watts (W), named after James Watt, who first converted the power of heat into mechanical power in a steam engine.

What can we measure on a wave that corresponds to steady voltage and current? If we measure the **peak** wave value of voltage and current for a sinewave (Figure 4.10) we find that the power figure we get by multiplying these quantities is twice as large as it would be if V and I were steady values. In symbols, $V_p I_p = 2W$, where W is the power value for steady voltage and current. This is not exactly surprising, because it is obvious that multiplying the peak value of

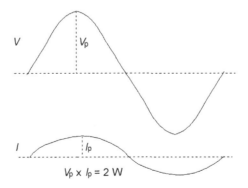

Figure 4.10:
Signal power. For a wave, using peak values of V and I gives a power that is two times too large. We therefore use RMS values such as $V_p/\sqrt{2}$ and $I_p/\sqrt{2}$ which will give the correct amount when multiplied

these varying values must give a result that is more than you would obtain by multiplying average values.

This is equivalent to using V and I quantities that are $V_p/\sqrt{2}$ and $I_p/\sqrt{2}$, respectively; if you multiply these quantities together you get $V_pI_p/2$, which is the true power in watts. These $V_p/\sqrt{2}$ and $I_p/\sqrt{2}$ quantities are called **root mean square (RMS)** values. We will omit the mathematical theory, but the name comes from the fact that these quantities represent the root of the average (mean) of the square of the peak quantity.

The important point is that if we want to make calculations on the power dissipated by a sinewave, we have to use these RMS quantities. For example, if the peak values of a sinewave are 6 V and 1 A, then the power is not 6 W but only 3 W. This used to be the basis of inflated figures for audio amplifier power outputs, with some manufacturers quoting real RMS power figures and other quoting peak power values (not divided by 2) or even rather imaginary values called **music power**.

■ Note

All of this assumes that the waves of current and voltage are in phase. If they are not, the amount of true power is calculated by multiplying the RMS voltage by the RMS current and multiplying also by the cosine of the phase angle. If the phase angle is 90°, then the power is zero.

■

Just in case you thought all this was looking quite logical and orderly, the figure of $\sqrt{2}$ is true only if the waveshape is that of a sinewave, and different factors (called **form factors**) have to be used if the waveform is different, as sound waves usually are. This, however, is a worry more for the designers of test equipment than for students of electronics. Fortunately, there are instruments that can measure the RMS values for any form of wave so that we do not need to depend on making peak measurements and performing elaborate mathematics.

Summary

For sinewave signals, the true power dissipated in a load can be found by multiplying together the RMS values of voltage and current, assuming that these waves are in phase. The RMS value is equal to the peak value for a sinewave divided by $\sqrt{2}$, a factor of 1.414. For example, the RMS value of a signal which is 10 V peak is $10/1.414 = 7.07$ V.

Feedback

A circuit method that is very important for linear amplifiers is called **feedback**. Feedback means taking a fraction of the output wave of an amplifier and connecting it to the input. This

Figure 4.11:
Feedback, positive and negative. The feedback signal is a fraction of the output that is added
to the input. If this signal is in phase with the input the feedback is positive; if it
is in opposite phase (anti-phase) the feedback is negative

fed-back signal can be connected so that it either adds to the normal input or subtracts from it (Figure 4.11). If the feedback signal is in phase and so is added to the input wave, this is called **positive feedback,** and its action is to increase the apparent gain of the amplifier, reduce its bandwidth, and make the gain figure of the amplifier more sensitive to any changes (such as a small change in the resistance value of a load resistor). If positive feedback is increased to the point where it is enough to provide all the input that the amplifier needs, the result is **oscillation**, where the amplifier will provide a wave output without any input.

Positive feedback is nowadays seldom used in normal amplifiers, but it is the basis of oscillators that are used to generate signals. Until positive feedback was discovered by Armstrong in 1912, radio signals were generated by rotating machines (alternators) and this limited the frequency of signals that could be used. When active devices are used in an oscillating circuit, the range of frequencies that can be generated is limited mainly by the design of the active components, so that it became easy to generate signals at frequencies of 1 MHz and more. Originally, Armstrong's positive feedback was also used to increase the gain of primitive radio receivers, but this type of use was abandoned in the 1930s because of the interference that was caused when users trying to hear a remote station would cause the radio to oscillate and radiate their own signals.

Negative feedback uses feedback signals that are in opposite phase to the input, and it reduces overall gain. It also increases bandwidth, reduces non-linear behavior, and makes the gain of the amplifier less sensitive to changes in the components. This has made it a standard method that is used by circuit designers who need particularly linear response, stability, and wide bandwidth. It is particularly applicable to opamps (see Chapter 3) and it allows the gain of an amplifier to be set by the ratio of two resistors rather than from elaborate calculations using figures that may not be particularly reliable.

■ Note

Negative feedback can reduce some types of distortion, but it does not reduce slew-rate distortion, and can actually cause slew-rate distortion to increase.

Summary

Signal feedback is extensively used in electronics circuits to modify circuit action. If the gain of an amplifier is more than unity, feeding back a portion of the output in phase to the input will increase gain, reduce linearity, and make the amplifier unstable, causing oscillation if the feedback is sufficient. Negative feedback will reduce gain, increase linearity, and make the amplifier more stable, provided the phase of the feedback signal remains at 180°. The performance of a negative-feedback amplifier is determined more by the passive components than by the active components.

Oscillators and Multipliers

An oscillator is a circuit that is designed to generate wave signals from a steady voltage supply, with no wave input. Any oscillator can be thought of as an amplifier with positive feedback and some type of circuit to determine the frequency. If this frequency-determining circuit is a resonant circuit the oscillator is a **tuned oscillator** and it will generate sinewaves, or at least waves that are close to a sinewave shape. Other circuits can be used that will cause the oscillator to generate a square or pulse waveform; these are **untuned** or **aperiodic** oscillators. Oscillators can also be designed so that their oscillating frequency can be varied when a steady voltage input is changed, and this type of oscillator is called a voltage-controlled oscillator (VCO).

A tuned circuit can also be used as a **frequency multiplier**. Suppose, for example, that an input signal at 1 MHz is applied to an amplifier which is not linear. The effect of the non-linearity is to change the waveshape, and this means that the signal will now contain other frequencies that are multiples of the original. In this example, the output of the amplifier will contain the 1 MHz signals along with others at 2 MHz, 3 MHz, 4 MHz, and so on. If the output stage is tuned to 2 MHz, this will become the predominant signal at the output, so that the effect of this stage is of a frequency-doubler, converting a 1 MHz signal into a 2 MHz signal.

CHAPTER 5

Block and Circuit Diagrams

Diagrams

A diagram is a picture that replaces, or supplements, words as a description of something. One obvious type of diagram is the one that is used for flat-pack furniture, consisting of a set of drawings of the components at each stage of assembly. At a time when radio technology was the main part of electronics, books and magazines for the amateur used to contain similar diagrams (Figure 5.1). This is a **component layout diagram**, showing where the components were fixed and how wires were connected between them. This example is from 1932 and it shows a constructor's diagram for a radio which was, for that time, a fairly advanced design (my father built it).

■ **Note**

A diagram like this of a simple circuit is very valuable if you are a beginner, because it shows in great detail how everything fits together. It is also very valuable for fault-finding, because each component can be located quickly. It does not, however, tell you much about how the circuit works.

■

The construction of electronics circuits has changed a lot since these days, but diagrams like this are still used, particularly for the home constructor or in servicing manuals so that the precise location of each component can be found quickly. The main difference is that layout diagrams are now simply an aid to location and are not used as the only information on an electronic circuit.

The trouble with component layout diagrams is that they specify components of a fixed shape. You could not use the diagram of Figure 5.1, for example, if every component you used had a quite different shape and size, particularly if you did not really know what you were doing. You could not use this type of diagram if you wanted to construct the circuit in a different physical form. You might, for example, want to construct a circuit that had to fit into a space that had been used for a different circuit. In addition, such diagrams have to be drawn by an artist, making them expensive to create, unlike circuit diagrams that can use computer-assisted methods. Worst of all, the diagram is no help in understanding what the circuit *does*.

Electronics Simplified. DOI: 10.1016/B978-0-08-097063-9.10005-6

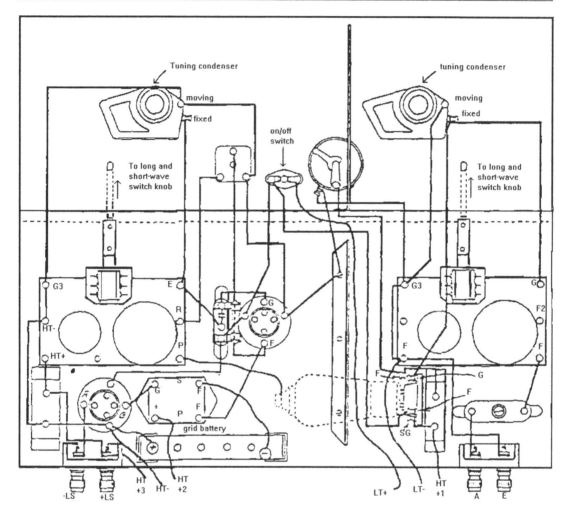

Figure 5.1:
A layout diagram of the type that was used by radio constructors in the 1930s. The diagram is useful only if you are using components that correspond to the shapes shown here

■ Note

Computer-assisted drawing has come a long way, and the layout of circuits on to printed circuit boards (see below) can be done by computer using fairly inexpensive software, enormously reducing the amount of cost and human effort that is needed for this task.

■

The modern equivalent of the old type of layout diagram is the **printed circuit board (PCB) diagram**. This shows how components are arranged on the PCB, which contains all the connections. The PCB is a strip of insulating material that is drilled so that the pins or wire

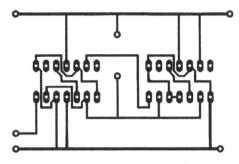

Figure 5.2:
A very simple printed circuit board, viewed from the connections side

connections of components can pass through, leaving the components lying close to the insulated board on one side, the component side. On the other side, the connection side (Figure 5.2), metal strips make the connections between components, and the pins or wires of the components are soldered to these metal strips. The important points are that the boards can be mass produced, with all the metal strips, using a photographic type of process, and that all the soldering can be done in one process by dipping the connections side of the board into a tank of molten solder that is agitated by high-frequency sound waves (ultrasonic waves).

The advantage of this type of construction is that all circuits on any particular design of PCB are identical and, because the placing of components and soldering can be automated, these assemblies can be produced at low cost, even if the assembly contains a large number of components. The testing of a complete board can also be automated. Figure 5.3 shows a typical PCB layout drawing that shows the component side of the board. Some such diagrams also show, in shaded form, the metal strips that connect the components on the other side of the board.

■ **Note**

PCBs can be made in rigid or flexible sheets of insulator. Though the rigid form is more familiar, the flexible boards can be used wherever the size or the shape of the space in which the board is to be fitted would not permit the use of a rigid board.

■

For some purposes, a single PCB is not sufficient and provision has to be made for additions to the board by plugging in additional boards. This is a typical form of computer construction, with the main board referred to as the **motherboard** and the plug-in units as **daughter boards**. Another type is the backplane, which might be called an inadequate-motherboard, as it consists mainly of interconnections for daughter boards rather than having a serious amount of circuitry of its own.

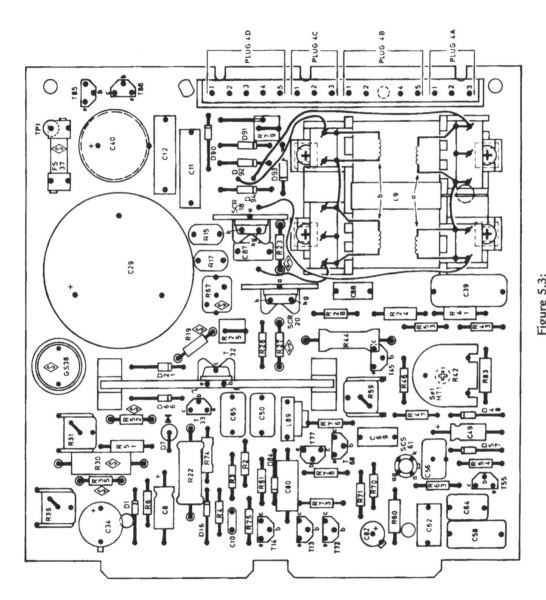

Figure 5.3:

A printed circuit board layout diagram, part of a Pye color television of 1978 vintage

Circuit Diagrams

A much more useful type of diagram, both for construction and for design purposes, is one that puts more emphasis on the *connections* between components rather than on the physical arrangement of components. This type of diagram is a **circuit diagram**, and Figure 5.4 shows an example taken from a book published in 1973.

In this type of diagram, components such as transistors or integrated circuits (ICs), resistors, inductors, and capacitors are represented by their standard symbols, and the connections between them are indicated by lines. In this old diagram, lines that are not connected are shown by a loop-over, but more recent diagrams show lines that cross at right angles to mean that no connections are made between these lines. In modern diagrams, all connecting lines are

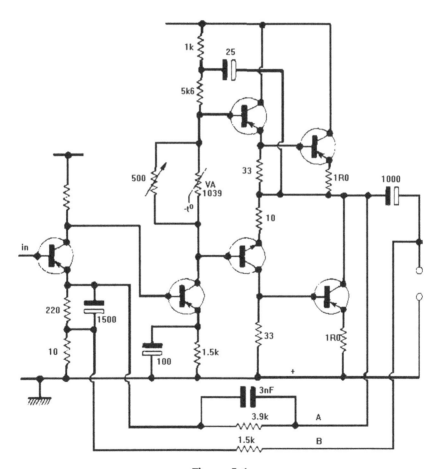

Figure 5.4:
A typical circuit diagram of the 1970s, showing the connections of the circuit. Components are represented by symbols and connections by joining lines. This form of diagram allows you to use whatever physical layout you prefer for construction

indicated by T-junctions, usually marked by a dot. Components are identified by using letters such as C, R, or IC (for capacitor, resistor, or IC) along with a number, so that a table of values can be used to find component values. Alternatively, as Figure 5.4 shows, the values could be printed on the diagram itself.

A circuit diagram is not such a useful guide to how a circuit will look when the components are put into place and connected up. It does, however, show the connections and how signal and steady currents will flow, so that after a little experience you find such a diagram much more useful than the layout type. For one thing, it can show you how the circuit works; for another, it allows you to construct the circuit in any form you want to use, not simply the fixed pattern that the constructional diagram forces upon you. The circuit diagram is the type that the designer and the constructor will work from. The layout diagram is still useful for servicing, so that you can quickly locate a component which, from looking at the circuit diagram, you suspect might have suffered damage.

Summary

The arrangement of components in a circuit can be shown in a layout diagram, but this is not useful if you are more interested in how the circuit works and what it does. The circuit diagram or **schematic** is a more fundamental type which represents components by symbols and shows connections as lines. A circuit diagram shows the action of the circuit, and need bear no similarity to the layout diagram. The layout diagram is still important for servicing work.

The coming of ICs made all circuit diagrams look rather different. There is no point in showing the circuits **inside** the IC, even if you know what they are. ICs are represented in diagrams by squares or rectangles, with connections numbered or lettered. Figure 5.5 shows a typical diagram of the 1980s, taken from a standard textbook of the time, *Servicing Electronic Systems* (Sinclair & Lewis, now in a 2007 edition as *Electronic and Electrical Servicing* by Sinclair & Dunton). This diagram shows the IC as a rectangle with 24 connecting pins, and ordinary (**discrete**) components, passive or active, connected to these pins. The circuits inside the IC, if we could show them, would probably require several pages of closely packed diagrams. This takes us one step backwards in understanding, because you can no longer work out how the circuit behaves simply by looking at the circuit diagram; you need also to know what the IC does, in terms of what inputs it requires and what outputs it provides. A modern circuit diagram may use only one or two ICs, so that a layout diagram is superfluous, but it must be accompanied by some information on the ICs.

Summary

Circuit diagrams are very useful for designers and for servicing work, but for anyone without a specialized knowledge of electronics they do not contain the right type of information. In particular, looking at a circuit diagram, even of the older type, does not tell you about what a circuit achieves unless you already know a considerable amount about electronic circuits.

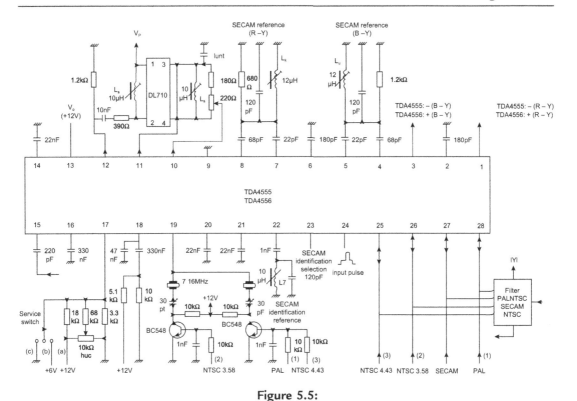

Figure 5.5:
A typical modern diagram in which the main component is an integrated circuit. The circuit paths are all inputs or outputs

There is just too much detail on this type of diagram, and the old adage about not seeing the wood for the trees applies. What you need to understand electronics **methods** is a diagram that concentrates on ends rather than on means, on signal paths rather than just the arrangement of components. Such a diagram is called a **block diagram**.

Block Diagrams

Definition

The shapes that you see in a block diagram are, not surprisingly (but not always), rectangular blocks. Each block shows signals in and signals out for a portion of a circuit, along with a name for the block and, very often, small sketches of the signal waveforms. All power supplies are ignored, as are individual components.

At one time, each block would have corresponded to a single IC, but for some block diagrams this would not be enough and several blocks may now be needed to explain the action of

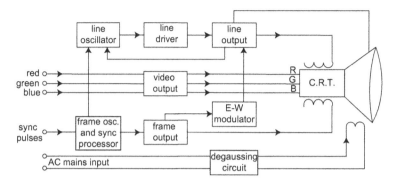

Figure 5.6:
A block diagram, in this example of a cathode-ray tube monitor for a computer

a single IC; after all, the whole of your circuit may be contained in one single IC nowadays. Figure 5.6 shows a typical simple block diagram for an old computer monitor using a cathode-ray tube (CRT).

Now you might think that Figure 5.6 is not a great advance in understanding. It is a block diagram for a computer monitor, and it brings home the point that no picture can completely replace words. If you are to make sense of a block diagram, you must know what the circuit is intended for and how it deals with the actions. The block diagram for a computer monitor does not mean a lot until you know what input signals exist and how they are used to affect a CRT. We will look at these aspects later, but the point is important: words and pictures go together and both are needed; you cannot learn from block diagrams alone. A picture, in the old proverb, may be worth a thousand words, but you still need a thousand words to explain a picture, even if it is not a piece of modern art.

Summary

A block diagram is used to understand quickly how a device works in terms of signal processing. The block diagram uses few symbol types and shows the effect of a circuit by sketching the input and output waveforms.

Linear Circuit Blocks

Figure 5.7 shows a simple linear circuit block diagram. In this example, the triangular shapes are used to mean linear amplifier circuits as an alternative to the rectangular block shape. If you think of the triangle shape as an arrowhead, it indicates the direction of signal from input(s) to output. The diagram is intended to show a simple audio amplifier that takes an input signal from a source (tape, CD, radio), amplifies the wave amplitude (voltage), applies volume and other controls, and then uses a power amplifier to produce an output to a loudspeaker.

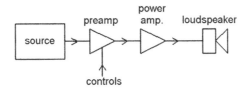

Figure 5.7:
A simple block diagram for an audio circuit

This illustrates the importance of words along with the diagram. Amplification, remember, really means making a copy of a waveform, and the copy usually is scaled up, with a greater amplitude of voltage. A circuit that carries out this action is a voltage amplifier, and an opamp is one example. Voltage amplification is always needed because the amplitude of signals from most sources like microphones, tape players, compact disc (CD) players, and so on is very small, anything from a few microvolts to a few millivolts. Once a wave has been amplified to a volt or so we can use controls such as volume controls, bass and treble controls, and so on. This part of an amplifier system is often called a **preamplifier**.

Why not use these controls on the smaller amplitude signals? The reason is **noise**. A signal is caused by the movement of electrons, but electrons are always in some kind of movement, making random shuffling motions in any conducting material: the hotter the material the more the electrons vibrate (and that is what we mean by **temperature**). If we connected an amplifier to any piece of conducting material and increased the amount of amplification (gain), the output would be a sort of signal, a jagged shape with no repeating pattern that we call electrical noise. When this signal is applied to a loudspeaker it sounds like a rushing noise, and when it is applied to a television screen it looks like white speckles on the picture.

Volume controls make rather more noise than other passive components, and we always try to avoid placing volume controls right at the start of an amplifier, because then the noise that they create would be amplified. If we amplify up the feeble signal that we want, and **then** use the volume control, we can ensure that the noise of the volume control is much less than the (wanted) signal amplitude. In electronics talk, we have a large **signal-to-noise ratio** (written as S/N, and measured in decibels). Where a large amount of linear amplification is needed we use a voltage amplifier (a **preamplifier**) first, one that is designed to add as little noise as possible. We can then use our volume and other controls on this amplified signal.

Linear ICs have steadily developed over the years, and it would be unusual nowadays to find separate preamplifier and power ICs. Accordingly, it is less usual to see the triangular symbol used on a block diagram because of the number of inputs and outputs. The main input, of course, is the feeble signal from the tape-head, vinyl disc pickup, microphone, CD reader, or whatever, but there are also connections for volume and other controls.

At one time, these would be signal outputs and inputs that could be connected to a variable potentiometer, but on modern circuits the control is carried out using a transistor circuit which uses a DC input. In other words, the higher the steady voltage applied to the circuit, the more signal it passes. This system ensures that the signal is processed entirely inside the IC, so that there is less risk of interfering signals being picked up at the control terminals, or of noise from the controls. This does not mean that the noise problem is any less, because there will inevitably be some noise from the internal circuits that carry out the volume control action. Active components are always likely to generate more noise than passive components, but the noise is easier to deal with than the noise from a volume control.

The point about block diagrams, however, is that they need not show too much detail. If you want to show a block for a circuit that carries out voltage amplification, you simply draw a triangle, show input and output points, and indicate what it does, using lines to show that there is provision for volume or other controls (Figure 5.8). More information is usually needed, and for a voltage amplifier, for example, you will need to know what amount of amplification (or **voltage gain**), expressed in decibels, is being achieved.

Summary

Block diagrams omit all the detail of a circuit diagram, and bear no relation to component layout. They are concerned with the actions that are performed on signals, and the usual system is to use one block for each action. Where this would lead to a diagram being too large, a set of actions can be represented by a block. The block diagram must show the inputs and outputs for each block, and this is often supplemented by other information such as a name and by waveform or timing diagrams.

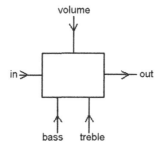

Figure 5.8:
A way of representing control inputs in a block diagram. A triangle amplifier symbol could be used here in place of the square illustrated here

Filtering

There is one important action that is very commonly used in linear circuits and which often has to be represented on block diagrams. **Filtering** means selecting one range of frequencies from others when there is a mixture of different signals present. You might, for example, want to filter a mixed signal so that it contained only the frequencies that your ear would detect if the signal were used to operate a loudspeaker or headphones. You might want to filter a mixed signal so that it contained only the higher frequencies, free from power-line frequencies. You might also want to filter a mixed signal so that you removed all the higher frequencies.

Definition

A filter is an electronic circuit that acts selectively on one frequency or a range of frequencies, either to pass or to reject that frequency or range.

When a filter passes only the lower range of the frequencies that are applied to it, we call it a **low-pass filter**. A typical graph of output plotted against input for such a filter is illustrated in Figure 5.9(a). A filter like this can be used in an audio amplifier to prevent unwanted signals at higher frequencies (**ultrasonic** frequencies) from reaching a loudspeaker.

You can also illustrate a graph for the response of a **high-pass filter** (Figure 5.9b), which rejects the lower frequencies of a mixture and passes only the higher frequencies. This can be used in an audio amplifier to prevent unwanted low-frequency noises (e.g. rumble from a turntable) from affecting the loudspeakers.

The other basic type of filter is the **band-pass type** (Figure 5.9c), which rejects both the highest and the lowest frequencies and passes only a range between these extremes. A radio tuner carries out this type of band-pass action so as to receive only one station and not a garbled mixture of all the signals it can pick up.

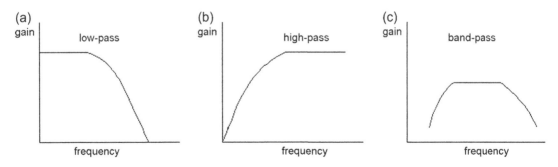

Figure 5.9:
Gain plotted against frequency (characteristic) graphs for three basic types of filters

Figure 5.10:
The symbols that are used to mark a filter in a block diagram

■ Note

You can also use a **band-stop** filter to reject a range of frequencies, and another possibility is a **spot** filter (or **sharp-cut** filter) which rejects one particular frequency along with a small range of frequencies around it. Older filter designs are analog, but digital filters can be designed for digital signals. Such digital filters often provide much more efficient filtering than the analog variety.

■

This sort of illustration is too much for a block diagram (though you might want to include it as part of the text explaining the block diagram). On a block diagram, we use a set of wave symbols to illustrate what a filter does. The single wave means low frequencies, the double wave represents middle frequencies, and the triple wave represents high frequencies. By placing a diagonal bar across a wave symbol we mean that the filter rejects this range of frequencies. Figure 5.10 shows some filter symbols used to indicate the basic types of filters.

Summary

Filters are important circuit blocks that are used when a frequency or range of frequencies must be singled out for use. The most common filter types are low-pass, high-pass and band-pass, but band-stop and sharp-cut filters are also used.

Digital Circuit Blocks

Digital circuit blocks make use of the same type of rectangular shapes as can be used for the more elaborate linear circuits. The action of the circuits is likely to be very different; for example, we are very seldom interested in amplification, but nearly always concerned with timing. The symbols that we show along with the block shapes are therefore very different.

Take a look first at a block diagram (Figure 5.11) for a computer, the standard diagram that was used for many years in (UK) City & Guilds 224 examinations for Electronic Servicing courses. Each block is of the same size on this diagram, and the names indicate the main parts of the computer (this is the main computer, not including items such as the keyboard, printer, or monitor). The main difference between this and linear circuit block diagrams (or between this any many other types of digital circuit block diagrams) is that there is no clear and obvious

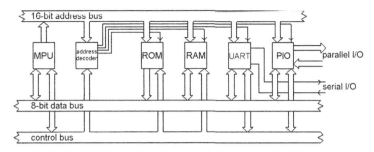

Figure 5.11:
An early computer block diagram, showing the representation of buses. This diagram will appear again in Chapter 13

input and output. We will look at this point (and this block) again in Chapter 13, but first of all, note the use of arrows.

The arrows show the direction of signals, and where you see double-headed arrows this means that signals will be passing in opposite directions, usually at different times or on different lines. The fact that these are broad (or block) arrows is also significant. When we want to indicate a single signal line, we can use a thin line with an arrowhead; there are some in this example. The broad outline arrows are used to indicate a set of signals. For example, we may have a set of 32 signals used as a signal connection to a block. Rather than show the individual lines (which goes against the principle of using a block diagram to eliminate detail) we simply draw a 'fat arrow'. When these lines are being used at one instant to provide input signals and at another instant to provide outputs, we indicate this by using the double arrowheads, one at each end of the thick arrow.

Even if you know absolutely nothing about a computer you can see from this diagram that, for example, the block marked MPU will send signals to the part called 'Address bus', and will both send to and receive from the part called 'Data bus'. A **bus** is, in fact, a collection of lines that can pass signals in either direction, so that when any block sends signals to a bus, anything else connected to that bus can make use of the signals. This is a type of connection that is used to a considerable extent in digital circuits to make connections simpler.

Let's look now at something that is lower down in the complication scale. Figure 5.12 shows a block diagram for a logic probe, an instrument that is used for trouble-shooting digital circuits. The principle is to connect a probe-lead to a line in a digital circuit so that the logic probe can indicate whether that line is at a low voltage (usually 0 V), at a high voltage (meaning +5 V or +12 V), or carrying pulses.

In this diagram, the input connection is indicated by the line labeled **probe**. The signal on this line is the input to a block marked **amplifier**, but this is not a voltage amplifier in the sense that we would use in a linear circuit. This amplifier is an example of the type of circuit that is called

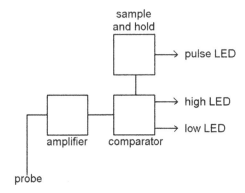

a **buffer**. It ensures that the power that is needed by the instrument is not taken from the circuit that is being tested. The voltage at the test point is the input to this amplifier, and the voltage out is the same. Whatever power is needed at the output is supplied by the amplifier, not by the circuit that is being tested.

The next block is labeled **comparator**. This compares the voltage from the amplifier (buffer) to find if this is high or low (the meaning of high and low will be clearer after you have read Chapter 9). If the sampled voltage is high, one of the outputs from the comparator will activate a light, the 'High' light-emitting diode (LED). If the output is low, the other main output is used to activate the 'Low' LED. If the output is pulsing, changing between high and low, this output is taken to the block marked **sample and hold**, and used to light a third LED.

Summary

Block diagrams for digital circuits use the same block shapes, but the connections are usually multiple lines called buses. The precise waveforms are not important, but timing is, and very often it is impossible to show the details of timing on the block diagram. An explanation in words is essential, and for many types of digital diagrams a timing diagram is also needed.

Other Blocks

Everything in a block diagram can be represented by blocks, but it is often convenient to add a few standard circuit symbols that can be recognized. For example, if a linear circuit drives a loudspeaker, the block diagram can show the standard loudspeaker circuit symbol because this is easy to recognize, and your eye will pick it up better than a box marked L/S. Other symbols that are often used are for a CRT, a motor, or a potentiometer, and some of these are illustrated in Figure 5.13.

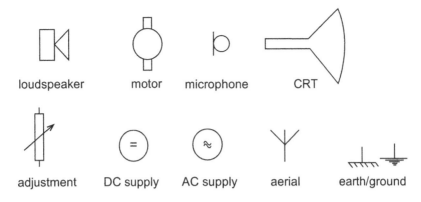

Figure 5.13:
Some symbols that are used along with block representations

■ Note

For an excellent set of illustrations of symbols, along with links to other useful sites, take a look at the website: http://www.kpsec.freeuk.com/symbol.htm

■

CHAPTER 6

How Traditional Radio Works

Radio Waves

We use the phrase **radio waves** to mean electromagnetic waves that are transmitted across space (not just through air), as distinct from waves conducted along wires or in tubes (**waveguides**). The existence of radio waves was predicted by the mathematician Clerk Maxwell in 1864 and later discovered experimentally by Heinrich Hertz in 1887, but no use was made of the waves until early in the twentieth century when Marconi (along with Popov, Tesla, and several others) used them for communications. They have come a long way in a period of about 100 years.

■ **Note**

This is a very familiar pattern: a mathematician discovers relations between physical quantities and years later we find that we can make use of these relations. Nations that do not nurture their mathematicians (on the grounds that the work seems pointless), or pretend that 'maffs' means just simple arithmetic, are doomed to be second rate.

■

We now know fairly well what happens. When we cause a voltage to exist between any two points in space, the **space** itself is distorted, and we refer to the effect as an **electric field**. A varying voltage causes a variable distortion which appears to us as magnetism; we say that a **magnetic field** exists, meaning that this piece of space causes magnetic effects. This is a changing magnetic field, however, and its effect is to generate another voltage some distance away, which in turn causes a magnetic field and yet another voltage field, and so on. The unique feature of the waves is that the direction of oscillation of the magnetic wave is at right angles to the direction of oscillation of the electric waves (Figure 6.1).

These alternating fields are waves of both electricity and magnetism, called **electromagnetic** waves. They travel away from the point where they were generated at a speed of around 300,000,000 meters per second (often written as $3 \times 10^8\,\mathrm{m\,s^{-1}}$), which is also the speed that has been measured for light, a strong hint that light is itself an electromagnetic wave. No air or any other material is needed to transmit these electromagnetic waves, though they can pass through insulating materials and they can also be conducted along metals and other conducting materials and through tubes called **waveguides**. Electromagnetic waves travel in straight lines,

Electronics Simplified. DOI: 10.1016/B978-0-08-097063-9.10006-8

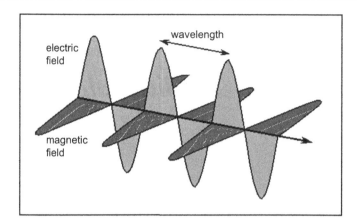

Figure 6.1:
Illustration of how electric and magnetic waves combine to form an electromagnetic wave

they can be reflected, refracted, and focused, and they can also be polarized (see later); all effects that are familiar from our knowledge of how light behaves.

In the early days of radio, all academic experts believed that long-distance communication by electromagnetic waves was impossible because of the curvature of the Earth. They would have known better if they had followed the work of Oliver Heaviside, who worked out that the effect of the radiation from the Sun would be to split atoms of gas in the upper atmosphere (the **ionosphere**) and provide a conducting layer that would reflect waves. Heaviside, however, was regarded as an eccentric engineer and was scorned by academics (does that sound familiar?), and much of his work was not published until after his death. Marconi and others disregarded the academics and pressed on with radio contacts over greater and greater distances, culminating in the transatlantic transmission in 1901. Academics then had to rediscover all that Heaviside had done and pretend that they had discovered it themselves. We will come back to the effects of the ionosphere later.

All electromagnetic waves travel at the same speed in space, and the speed is almost the same in air, though slower in denser materials. In addition, these waves have a measurable **wavelength**, meaning the distance from the peak of one wave to the peak of the next. As for water waves and sound waves, the speed, frequency, and wavelength are related by the equations:

$$v = f\lambda \text{ or } f = v/\lambda \text{ or } \lambda = v/f$$

with f meaning frequency in hertz (Hz), v speed in meters per second (m s^{-1}), and λ wavelength in meters (m). For example, if you generate a signal at 1 MHz, then its wavelength in space is 300,000,000 divided by 1,000,000, which is 300 m. A wave at 100 MHz, such as we use for FM radio, has a wavelength of 300,000,000/100,000,000 which is 3 m. Table 6.1 shows a few examples of frequencies and wavelengths.

Table 6.1: Wavelength and frequency.

Frequency	Wavelength
100 kHz	3 km
1 MHz	300 m
5 MHz	60 m
10 MHz	30 m
20 MHz	15 m
50 MHz	6 m
100 MHz	3 m
400 MHz	75 cm
1 GHz	30 cm
10 GHz	3 cm

Definition

Radio waves are electromagnetic waves in space (which includes air) traveling at around 300 million meters per second. The frequency of a wave is the number of oscillations per second, and this causes a wave to have a wavelength which is found from speed/frequency.

■ Note

Electromagnetic waves travel at a lower speed in insulating materials, and also when they travel along wires or in metal tubes. The relationship between wavelength, frequency, and speeds is the same, but waves traveling in insulators or along wires have a shorter wavelength than the same waves in space because of the lower speed. ■

It is not difficult to generate and launch waves in the frequency range that we use for radio. Anything that produces an alternating voltage will generate the signals, and any piece of wire will allow them to radiate into space. We find, however, that a launching wire, which we call an antenna (aerial in the UK), works best when its length is an even fraction (such as $\frac{1}{2}$ or $\frac{1}{4}$) of the wavelength of the wave, or several wavelengths long. Lengths of half a wavelength or quarter of a wavelength are particularly useful, and a very efficient system, called a **dipole**, has been used for many years. These tuned aerials are not so efficient (and are difficult to construct) for the very long wavelengths, and they really come into their own for wavelengths of a few meters or less. Though we still make considerable use of the very long wavelengths, most of the communications equipment (FM radio, television, mobile phones) that we use nowadays depends on using the short wavelengths. The limitations of short wavelengths for long-distance communications are overcome by using satellites, so allowing a straight-line path for signals to and from the satellite.

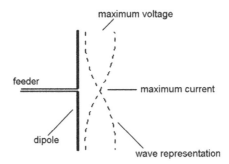

maximum voltage

feeder

maximum current

dipole

wave representation

Figure 6.2:
Dipole antenna principles. The dashed lines represent the half-wave of signal at the antenna, ensuring the maximum signal current to the feeder line of a receiver (or from the feeder of a transmitter)

A dipole antenna (Figure 6.2) for a wavelength that you want to use consists of two rod sections, each one-quarter of a wavelength long. If you imagine the inside portions connected to a signal, the voltage on the antenna rods will be in the pattern of a wave, and when the voltage is almost zero at the center it will be a maximum at the ends. This is the type of antenna that is very familiar to us and is used for television and FM radio reception, with the rods cut to the correct quarter-wavelength size for the nearest set of transmitters. The dipole is most efficient for the wave that is exactly of the size, but this is not too critical, particularly if the rods of the dipole are thick. The points where the (**feeder**) cable is attached correspond to the points of maximum signal current so that the antenna behaves like a low resistance as far as signals of the tuned frequency are concerned.

Dipoles are also polarized. Polarization, as far as radio antennae are concerned, means the direction of the *electric field* of the wave, and this is often fixed by the design of the transmitting antenna as either vertical or horizontal. If the transmitter uses a vertical dipole, the signal will be vertically polarized, and the receiver antenna should also be vertically polarized. If the transmitter uses horizontal polarization, the receiver antenna should also be horizontally polarized. Different polarization directions are used to reduce interference between transmitters that use the same (or close to the same) frequency, and in some cases because one direction of polarization provides better reception conditions in difficult terrain than the other.

▪ Note

Where the length of a dipole might be excessive, or if a dipole would be inconvenient (as for a car phone, for example), we can use the upper half of a dipole, often referred to as a **monopole** or **whip antenna**. The length of a dipole can be reduced by wrapping a wire dipole around a magnetic material, and this applies also to a monopole, as used for mobile phones.

Summary

Radio waves all travel at the same speed in space, and have measurable wavelength and frequency values. These quantities are related, and speed = wavelength × frequency. Radio waves are of the type called electromagnetic waves, differing in frequency, and including light, infrared and ultraviolet, X-rays, gamma-rays, and many others that were discovered but not originally thought to be electromagnetic. A radio wave is most efficiently radiated from a metal antenna whose length is a suitable fraction of a wavelength, such as half-wave, and reception of radio waves is best when the receiving antenna is also a suitable fraction of a wavelength. For low frequencies with wavelengths of several kilometers, half-wave antennae are impossible, and it is fortunate that half-wave antennae are most needed for waves whose wavelengths are less than a couple of meters. Dipole antennae are polarized, and their direction of polarization should match that of the transmitter.

Transmission and Reception

Early radio transmitters used electrical sparks to generate signals, which were a mixture of many frequencies, mainly in the very high-frequency (VHF)/ultrahigh-frequency (UHF) range, and at a short range these were seen to cause sparks to appear across the two sections of a receiving antenna. More sensitive ways of detecting the received signal soon followed, and a device called a coherer, consisting of metal filings in a glass tube, was used extensively, along with headphones, in the early days of radio as a way of detecting signals that were too feeble to make a spark.

For some time, large land-based transmitters used mechanical generators, alternators, working at about 20 kHz, and with huge long-wire antennae, so that their signals could span the Atlantic. These, and spark generators, were used until radio vacuum tubes were invented around 1906, though spark transmitters were still in use when the Titanic sank in 1912. An interesting point is that her distress signals were picked up in New York by a Marconi radio operator called David Sarnoff (who later founded the Radio Corporation of America when the US government threatened Marconi's with the equivalent of nationalization).

Reception of radio waves is simple enough: make a dipole or a single-wire antenna, and detect the small currents at the center terminals. Detection is a snag, though. At the start of the twentieth century, there was nothing that could amplify a signal at 20 kHz, and devices such as earphones could not respond to such higher frequencies or the much higher frequencies that were generated from sparks. The only way that the presence of the wave could be detected was by what we now call **rectification**, using some device that would allow the current to flow only one way and then detect the direct current (DC). The coherer with its metal filings was the first attempt to make such a device, but later it was discovered that crystals of the mineral galena (chemically, lead sulfide) could be used. The crystal was placed in a metal holder connected to

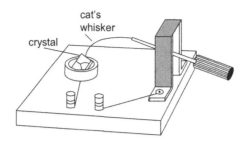

Figure 6.3:
A crystal set. This vintage radio diagram shows the use of a crystal and a fine wire (cat's whisker) to rectify radio waves into the audio signal. The resulting signals could be heard in earphones. A handle allowed the whisker to be moved on the crystal surface

earphones, and an antenna was connected to a fine wire, called a **cat's whisker**, jabbed against the crystal (Figure 6.3).

When a radio wave hit the antenna, the alternating current flowed only one way through the coherer or the crystal, and the headphones gave a feeble click, which sounded reasonably loud if you happened to be wearing the headphones. If the radio waves had been switched on and off again quickly, the sound in the headphones was a short double-click; if the wave was sustained for longer the gap between clicks was longer. By switching the transmitter on and off using a Morse key, a skilled listener could read the short (dot) and long (dash) intervals between clicks and write down the Morse code message that was being transmitted. This was the basis of radio as we used it right into the 1920s.

Summary

Anything that can generate electrical oscillations can form the basis of a transmitter, and early designs used sparks, generating a huge mixture of wavelengths. Later designs used mechanical generators for low-frequency waves that required huge antennae. Communication was carried out by using Morse code, switching the signals on and off in the familiar dot and dash patterns.

Neither coherer nor crystal was really satisfactory. The coherer had to be tapped regularly to settle the filings into place, and the cat's whisker had to be moved over the crystal surface to find a 'sweet spot', a place where the action was most effective. This movement of the cat's whisker often had to be repeated, usually just as you particularly wanted to get the important part of a message.

Modulation

Morse code served radio well and is still useful for emergency purposes because it requires the absolute minimum of equipment and it is the most efficient use of radio; even a low-power

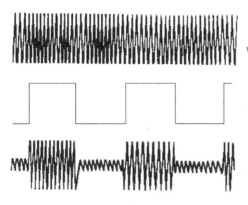

unmodulated carrier - a high-frequency sinewave whose amplitude is constant

square wave modulation

modulated carrier - the amplitude of the sinewave rises and falls in the pattern of the square wave and at the same frequency

Figure 6.4:
Amplitude modulation. The signal, shown here as a square wave, alters the shape of the carrier outline so that the resulting outline (the envelope) has the same shape and frequency as the modulating wave

long-wave transmitter can have a very long range using Morse. It is not exactly useful for entertainment purposes, however, or for long-distance telephone calls, and though a method of carrying sound signals over radio waves had been demonstrated early on (on Christmas day in 1906), it took a long time to make this a commercial proposition (in 1920, in fact). The use of radio waves to carry the information of sound waves requires what we call **modulation**, and in the early days of radio this was always the type we call **amplitude modulation** (**AM**).

■ Note

It's an amusing thought that tapping out a message in Morse is still faster than trying to type a text message on a mobile phone, and it uses only a tiny fraction of the bandwidth.

■

To start with, we have to change the sounds, whether speech or music, into electrical waves, using a type of transducer called a **microphone**. Microphones were, thanks to the use of telephones, reasonably familiar devices in 1906 (in the USA at least), so that the only obstacle to broadcasting in these early days was knowing how to make a radio wave of 100 kHz or more carry a wave, the audio signal, of a much lower frequency, about 300 Hz. The simplest answer is to make the amplitude of the radio wave rise and fall in sympathy with the lower (audio) frequency (Figure 6.4). This is what we mean by **amplitude modulation**: the amplitude of the carrier wave is controlled by the amplitude of the audio signal.

Amplitude modulation had been possible even in the early days of spark transmitters, but only with difficulty and using primitive equipment. The use of radio **vacuum tubes**, or just 'tubes', in the USA, made efficient modulation possible, because the vacuum tube (called a valve in the

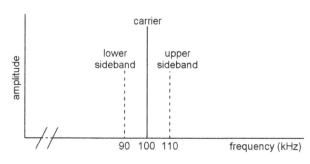

Figure 6.5:

Sidebands. A carrier can be represented as a vertical line on a graph of amplitude plotted against frequency. When this carrier is modulated by a sinewave, two more frequencies are present, the upper and lower sidebands. The transmitter must provide the extra power represented by these additional frequencies

UK) allows a large current to be controlled by a small voltage. In addition, another tube can be used as an oscillator, generating the radio waves and replacing the mechanical or spark type of generators. The first radio station intended for entertainment started operation in November 1920 at Pittsburg, Pennsylvania. The usual way of modulating was to control the DC supply to the output tube of the transmitter, using another tube, and applying the amplified audio signal as the input of this tube.

When a radio wave is modulated in this way, it is no longer a simple sinewave. If we could graph the waveform we would find the type of shape illustrated in Figure 6.4, but there is another effect that is shown only when we graph the signal amplitude against **frequency** rather than against time.

Suppose, for example, that we modulate a 100 kHz radio wave with a 10 kHz sinewave signal. Analyzing the frequencies that are present shows that three main waves now exist. One is at 100 kHz and is the **carrier wave**, the wave that is generated by an oscillator. There is also a wave with a frequency of 90 kHz, called the **lower sideband**, and one with a frequency of 110 kHz, called the **upper sideband**. These sidebands appear only when the carrier is being modulated, and their amplitude is always less than that of the carrier (Figure 6.5). In general, if the carrier frequency is F and the signal frequency is f, the sidebands are at $F - f$ and $F + f$, so that the total bandwidth of the transmitted wave is from $F - f$ to $F + f$, making a bandwidth of $2f$.

If we modulate the carrier with a signal that consists of a mixture of frequencies up to 10 kHz in bandwidth, the sidebands will show a pattern such as appears in Figure 6.6, showing a spread of sideband frequencies rather than just two single frequencies. In each sideband, the frequencies closest to the carrier frequencies are due to the lower frequencies of the modulating signal, and the frequencies furthest from the carrier frequency are due to the higher frequencies of the modulating signal.

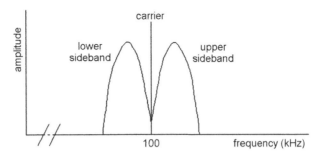

Figure 6.6:
Audio signal sidebands. When the carrier is modulated by a range of frequencies, as it would be for an audio signal, the sidebands cover a range of frequencies that is equal to the audio bandwidth. Both upper and lower sidebands contain the same signal information

All we need to carry the information is one of these sidebands, so that both the carrier frequency and the other sideband are redundant. There are ways that can be used of eliminating the carrier (**suppressed carrier** transmission) and one sideband (**single-sideband** transmission) from the transmitted signal, but for entertainment purposes the additional complications at the receiver make these methods undesirable, though both stereo radio and color television make some use of these methods for additional signals (using **sub-carriers**; see later). Digital radio is quite another matter, as discussed later.

Summary

Modulating a radio carrier wave always causes sidebands to appear. For amplitude modulation, the difference between a sideband frequency and the carrier frequency is equal to the modulating frequency. A modulated wave therefore requires a greater bandwidth than an unmodulated wave, and its efficiency is low, because the carrier and one sideband contribute nothing: all of the information is contained in each sideband (so that either sideband can be used to extract the signal at a receiver).

Reception

At the receiving end, vacuum tubes were still a far-off dream for the few people in 1920, mainly amateur enthusiasts, who could get hold of crystals and headphones and rig up an antenna. Some 1000 listeners probably heard the 1920 US presidential election results in this way, and in later years this number multiplied rapidly. Even in controlled and restricted Britain listeners experimented, trying to hear the trial transmissions from the Marconi workshop in Writtle (Essex) and even from continental broadcasts when conditions and geographical position suited. In 1922, a private company, the British Broadcasting Company (BBC), was formed and it set up nine transmitters, each with a distinctive call sign of digits and letters. Table 6.2 shows these nine with opening dates and frequency allocation.

Table 6.2: The nine transmitters of the BBC: opening dates and frequency allocation.

London	822 kHz	2LO
Birmingham	626 kHz	5IT
Manchester	794 kHz	2ZY
Newcastle upon Tyne	743 kHz	5NO
Cardiff	850 kHz	5WA
Glasgow	711 kHz	5SC
Aberdeen	606 kHz	2BD
Bournemouth	777 kHz	6BM
Sheffield	980 kHz	2FL

By the 1930s, radio as we know it was a reality. The BBC had been nationalized so that the government could control it and tax it, but in the USA broadcasting remained free in every sense and consequently the USA remained at the front of technical progress. Small tubes became available and were used in receivers, and by the 1930s loudspeakers were normally used in place of headphones so that the whole family could listen in. Amplitude modulation was used for transmissions with frequencies in the medium waveband of about 530 kHz to 1.3 MHz, and the receivers developed from the simple one-tube pattern to multitube types, some of the expensive models featuring an added electronic record-player to make a radiogram. By the late 1930s a few lucky people in the London area could, if they could afford it, buy a combined radio and television receiver (though television was confined to a couple of hours in the evening, and a receiver cost as much as did a small house at that time).

A severe problem with the use of AM and the medium waveband was interference between stations. Medium-wave broadcasts can be picked up over quite long distances, up to 1000 miles (over 600 km), and where two stations used frequencies that were close, they set up an interference that could be heard as a whistle at a receiver tuned to either station. The whistle, in fact, is at the frequency that is the difference between the two transmitted frequencies. For example, if one station transmits at 600 kHz and another at 605 kHz, the whistle is heard at a frequency of 5 kHz. As new radio transmitters were introduced all over the world, it became quite impossible to share out frequencies on the medium waveband, and interference became the common experience for all listeners, particularly after dark when the reflecting layers in the ionosphere moved higher and reflected waves that originated from greater distances.

A way of getting around these problems had been devised by an inventor who, of all that ill-treated fraternity, was the least fortunate. Edwin Armstrong developed **frequency modulation** (FM) to eliminate the effects of interference, both artificial and natural, and built a radio station (now a working museum) in the Catskill mountains just outside New York to prove the point. As the name suggests, the low-frequency audio signal is carried by altering the **frequency** of the radio signal (Figure 6.7), with the amplitude kept constant. Since interference of all types only alters the amplitude of radio waves, FM can be virtually free from interference unless it is

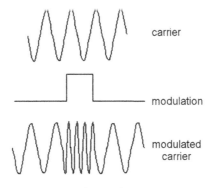

Figure 6.7:
Principle of frequency modulation. The amplitude of the carrier remains constant, and the signal alters the frequency of the carrier. The amount of alteration is called the deviation of the carrier. The bandwidth of sidebands created by this type of modulation is larger than that created by amplitude modulation

so severe as to reduce the radio wave amplitude to almost zero. FM is used universally today for high-quality broadcasts and for short-range local radio because if two FM transmitters are using almost the same frequency, a receiver will lock on to the stronger transmission, ignoring the other one and with no whistle effects. There is strong resistance to the proposal to turn off FM broadcasts all over Europe because of the poor coverage of digital radio (digital audio broadcasting or DAB) in so many parts of the UK. FM broadcasting in the UK did not start until 1955, though there were some commercial FM transmitters in the USA by 1941.

Summary

Medium-wave broadcasting became so popular that by the later 1930s there were too many transmitters to permit a reasonable bandwidth to be used by each, and the interference between transmitters that were on adjacent frequencies caused whistling noises in receivers. The solution was to use a different modulation system, FM, and to transmit FM signals using high-frequency carriers, typically in the 100 MHz region.

Radio Block Diagrams

Early Receivers

Early radio receivers for AM signals consisted only of an antenna, a crystal, and headphones. This arrangement could be used close to a transmitter, and when a coil and capacitor were added it became possible to tune to more than one transmitter signal (assuming there was more

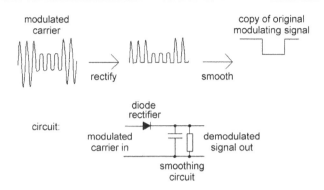

Figure 6.8:
Principles of amplitude demodulation (detection). The diode rectifier passes only one half of the modulated carrier, and filtering leaves only the outline which corresponds to the modulating signal

than one transmitter within the very limited range of a crystal receiver). The crystal is the **detector** or **demodulator** that allows the low-frequency signal to be extracted from the radio wave, and the principles of AM now are still the same as they were then. When a modulated signal is passed through any device that allows only one-way current (a **demodulator**), only half of the modulated wave will pass through. As Figure 6.8 shows, this makes the signal asymmetrical, and a small-value capacitor between this point and earth (together with the resistance of the earphones) will integrate the signal, ignoring the rise and fall of the radio frequency waves and following only the modulation signal.

All radio receivers have developed from this simple beginning, and the two aims of development have been to improve both **sensitivity** and **selectivity**. Sensitivity means the ability to pick up and use faint signals from remote or low-power transmitters. Selectivity means the ability to separate radio signals that are of closely adjacent frequencies. Both are important if you want to use a radio with a large choice of transmissions.

Sensitivity requires amplification, and selectivity requires tuning. Though this was well understood in the early days of radio, there were always problems. For example, if you amplify a radio wave too much there is a danger of **positive feedback** (when the amplified signal can affect the input), causing the receiver to oscillate, and drowning out reception for all other receivers near it. If a receiver is too selective, the sound that you hear is unintelligible, lacking the higher frequencies. In the phrase that was used at the time, it is a 'mellow bellow', the sort of noise you now hear from cruising cars.

By the start of the 1930s, a typical radio receiver followed a design called **TRF**, meaning tuned radio frequency, for which the block diagram is shown in Figure 6.9. The feeble signal from the antenna is both amplified and tuned in one or more amplifying stages that used tubes. The tuning circuit used a coil and a variable capacitor (two sets of metal vanes separated by air and arranged so that they could mesh in and out). If more than one tuned circuit was used for

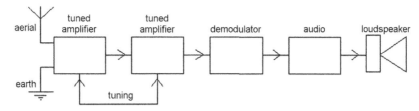

Figure 6.9:
A block diagram for a tuned radio-frequency (TRF) receiver, the first type of receiver that was constructed using vacuum tubes

greater selectivity, the tuning capacitors had to be 'ganged', meaning that they could be moved in step with each other, using a single metal shaft to carry all the moving vanes of all the capacitors, and a single control. The sensitivity of these radios was often improved by a small amount of positive feedback of the radio signal, and the crystal that had been used in the early days was replaced by another tube, a **diode**, that carried out the action of demodulation.

The action of the diode demodulator was the same as that of the older crystal, but with the advantage that tube diodes could be mass produced and were much more reliable because they did not rely on a contact between a wire and a crystal, all in open air. The output signal from the diode was still very feeble, more suited to earphones than to a loudspeaker, so that the natural path of development was to add another tube amplifier for the low-frequency audio signal, usually with a low-pass filter to get rid of the remains of the radio-frequency signal and so prevent it from being fed back to earlier stages.

These radios were a very considerable improvement, in both sensitivity and selectivity, on the old crystal sets, particularly when more than one stage of tuning was used, but as the medium waveband started to become crowded the old problems returned. Selectivity was still not enough, and attempts to increase sensitivity caused positive feedback and 'howling'. This latter problem was caused by excessive feedback that caused a receiver to oscillate and transmit an interfering frequency for other receivers. Causing howling was looked on as highly antisocial, and some remedy had to be found. Attempts to make radios with three or more tuned circuits made the problem worse, because with three tuning capacitors on one shaft, there was always a path for positive feedback of waves from the output to the input.

Summary

The most primitive radio system following the crystal set was the TRF receiver. A set of tuned circuits along with amplifier stages was used to select and amplify the wanted frequency. This amplified signal was then demodulated and the audio signal further amplified to drive a loudspeaker. The disadvantage was that the tuned circuits had to be ganged, which made it impossible to isolate them from each other, so that positive feedback and oscillation were always a problem.

The Superhet

The solution to the problems of medium-wave radio lay in yet another earlier invention by Edwin Armstrong. This bore the full title of the **supersonic heterodyne** receiver, abbreviated to **'superhet'**, and it is still the main type of circuit that we use for all reception of radio, television, and radar today. The principle is to eliminate as far as possible the amplification of the carrier frequency, so that variable tuning is used only at the start of the receiver block. The invention deliberately makes ingenious use of the 'whistle' frequency that is generated when two radio frequencies are mixed together.

The principle, as it is used in medium-wave radios, is illustrated in Figure 6.10. A tuned circuit selects an incoming carrier, or a small range of carrier frequencies, from the antenna and this is used as the input to an amplifier. At the same time, another frequency is generated in an **oscillator** circuit and applied to the same amplifier. In a conventional medium-wave receiver this generated frequency is not the same as the received frequency, it is exactly 550 kHz higher, and the tuning capacitor for the oscillator is ganged to the input tuning capacitor so that these frequencies stay exactly 550 kHz apart as the shaft of the capacitors is turned to change the tuning.

■ **Note**

The block diagram shows actions that were often combined in older radios, so that, for example, the tuned amplifier, oscillator, and mixer actions were often carried out by one radio tube. Later, actions were separated and each action was carried out by a separate transistor. Later still, the whole set of actions could be carried out in a single integrated circuit (IC).

■

The result is that the outputs from the first stage, called the **mixer**, consist of four lots of radio signals. Suppose, for example, that the incoming radio wave is at 700 kHz and is amplitude modulated. The oscillator will be set to generate an unmodulated 1250 kHz radio wave, and the

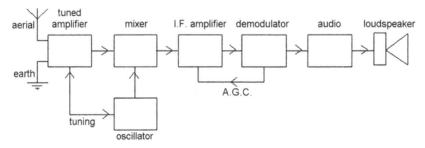

Figure 6.10:
A block diagram for a typical superhet receiver, showing the automatic gain control (AGC) connection

result is that the output of the mixer consists of waves of 700 kHz, 1250 kHz, 550 kHz, and 1950 kHz. These are the input signals plus the sum and difference of the frequencies. These sum and difference frequencies are modulated exactly like the input frequency.

Now the 550 kHz signal is easy to separate by a tuned filter, and it can be amplified. Any feedback of this signal to the input of the amplifier is not likely to cause much harm, because it is at a very different frequency from either the input wave or the generated wave. In addition, because this new frequency, called the **intermediate frequency** (**IF**), is fixed, it can be tuned by circuits that are fixed; there is no need to try to alter the tuning of these circuits when you tune from one station to another. As an extra precaution, metal boxes can be put over the IF tuned circuits to reduce any feedback even further. Adding more IF stages dramatically increases both selectivity (because there are more tuned circuits) and sensitivity (because there are more amplifier stages), so solving, for quite a long time, the problems of the crowded medium waves.

One feature that was used more and more, even in the early days, is **automatic gain control** (AGC). The superhet can be a very sensitive receiver, and if it is sensitive enough to provide a usable output from the faint, faraway, transmitters, then the nearby ones are likely to overload it, causing severe distortion. In addition, because radio waves are reflected from shifting layers of charged particles in the sky (the ionosphere), the received signal usually fluctuates in strength unless it comes from a nearby transmitter. Figure 6.11 illustrates this, showing a wave that can reach a receiver by two paths, one of which is a reflected path.

At any instant, these two waves can be in or out of phase. When they are perfectly in phase, they add so that the signal strength is increased compared to the strength of a single wave. When the waves are out of phase, the signal strength is reduced. Because the reflecting layers in the atmosphere are charged particles located a few hundred miles above the surface of the Earth and continually moving, the phase of the reflected signal is constantly changing, and so the received signal strength continually fluctuates.

This is less of a problem for FM radio and television signals, because the higher frequencies that these services use are not reflected to anything like the same extent by the layers in the upper atmosphere, and only the direct wave is used over a comparatively short range, of 100 miles (about 60 km) or less. Occasionally, a sunspot will greatly increase the number of charged particles in the atmosphere, and on the old analog television receivers you used to see interference on the screen resulting from the reception of distant transmitters. When the changeover to digital television is complete these problems will no longer exist unless sunspot activity is very severe.

■ **Note**

Very severe sunspot activity could have a drastic effect on all of our communications, including satellite signals.

■

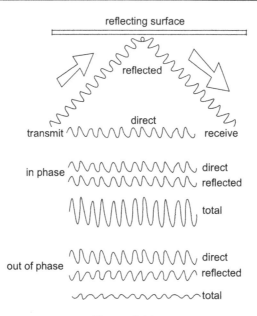

Figure 6.11:
Why automatic gain control (AGC) is needed. Unless a receiver is very close to the transmitter, a mixture of direct and reflected waves will be picked up, and the phase of the reflected waves will vary as the charged layers (ionosphere) move. This causes alternate reinforcement and reduction in total signal strength

At the demodulator of any radio, the effect of the diode is not just to extract the audio signal; there is also a steady voltage present. This voltage is obtained from the effect of the diode on the radio frequency or IF signal, and it is steady only while the strength of the incoming signal is steady; it is large for a strong signal and small for a weak one. The remedy for varying signal strength, then, is to use the steady voltage at the demodulator and feed it back to the IF amplifier stages. The tubes that were used for the IF amplifier and (usually) for the mixer stage were of a specialized type (called **variable mu**) in which the amount of amplification changed when the value of steady voltage applied to an input was changed. By making this feedback connection, the tubes could be made to work at full gain when the signal strength was low, but at reduced gain when the signal was large. Using AGC, the receiver could cope automatically with the changes and avoid fluctuations due to shifting reflections.

That name? The mixing of waves is called a **heterodyne** action, and the oscillator operates at a frequency that is **supersonic**, meaning higher than frequencies that we can hear (we started to use it to mean faster than sound much later). By the later 1930s you could hardly hold your head high in polite society unless your radio was a superhet. By 1939 the more elaborate radios would use eight tubes. One would be used to amplify the radio frequency and followed by a mixer stage to obtain the intermediate frequency, with two more stages of amplification for the IF signals. Following the demodulator there would typically be four tubes used for the audio (sound frequency) signals, two of them used to drive the loudspeaker.

At the bottom end of the scale, you could buy three-tube radios, with a single tube carrying out oscillator and mixer actions, one IF stage, and a combined demodulator and audio output value. All these tube counts would be increased by one for a set designed to be run from the AC mains, with this additional tube rectifying the AC to a one-way voltage and with a large capacitor added to filter out the remaining AC and smooth the fluctuating voltage. This was the type of radio I, along with thousands of others, built for myself in the late 1940s when it again became possible to buy radio components after the war.

■ Note

An alternative to the superhet was the **homodyne** receiver, in which the oscillator was at exactly the same frequency as the carrier. In such an AM receiver, the output of the mixer is the audio signal. The homodyne was never developed in the early days because of the difficulty of maintaining precise oscillator frequency, but the principle has been revived as a proposal for digital radio and television receivers, allowing simpler (and cheaper) circuitry for demodulating the digital signals. The new name is **zero-IF**.

In the mid-1950s, transistors started to replace tubes, but the block diagram remains exactly the same because the superhet principle has never been superseded for this type of radio use. The main changes in the years from 1960 to the present day have been the replacement of transistors by ICs and the increase in the use of FM radios, These changes do not appear on the block diagram, because even the use of FM mainly concerns using a different type of demodulator; the radios are still superhet types. The AGC principle could be applied even more easily to transistors than to tubes, but FM radios use, in addition, another type of automatic control, automatic frequency control (**AFC**). Superhet action also applies to radar and to digital television or radio. Let's be proud of Edwin Armstrong.

FM radio makes use of carriers in the higher radio frequencies in the 80–110 MHz range, and an IF of 10.7 MHz, and it was initially more difficult to make oscillator circuits that would produce an unchanging frequency in this range. Oscillators suffer from drift, meaning that as temperature changes, the oscillator frequency also changes. The percentage change might be small, and for an AM receiver working at 1 MHz, a 0.01% drift is only 100 Hz and not too noticeable, but for an FM oscillator working at 100 MHz a 0.01% drift is 10 kHz, totally out of tune. The FM signal needs a wider bandwidth than AM, and usually up to 230 kHz is allowed. Modern designs (as used in car FM radios and in portable receivers) use digital tuners that provide much better performance.

AFC is a must for FM receivers. This uses another steady voltage that is generated in an FM demodulator and which depends on the frequency of the signals. By using this voltage to control the frequency of the oscillator (a **voltage-controlled oscillator** or **VCO**), the FM receiver can be locked on to a signal and will stay in tune even if the oscillator components change value as they change temperature. Modern FM receivers have very efficient AFC which is usually combined with a muting action so that there is no sound output unless you are perfectly tuned to a transmitter. As you alter the tuning, each station comes in with a slight plop rather than with the rushing sound (of noise) that was so familiar with the earlier FM radios. The most modern FM receivers use digital tuners, meaning that a crystal is used to produce a stable frequency that is then used in a frequency-synthesizing IC to produce an output at any frequency in the set that can be used for FM reception. Digital tuners of this kind are almost universal on car radios and are available on some (notably Panasonic) portable FM radios.

The main changes in all radios since the 1970s have been the replacement of transistors by ICs so that modern radios contain very little circuitry and most of the space is used for the battery and the loudspeaker. The circuitry is almost identical on all of them, and different brands are often made in the same factories, which can be anywhere in the world. As we know so well, the name on a radio is no guide to where it was manufactured, and the manufacturing of consumer electronics in the UK virtually came to an end in the 1960s when a misguided attempt by the government to protect the electronics industry made it much cheaper to import complete circuits than to manufacture them from highly taxed components. This sort of thing has happened so often now that you might think politicians would have learned from it.

Summary

The superhet principle was one of the most important events in radio history, and is still in use. The incoming radio frequency is converted to a lower, fixed, intermediate frequency, and this lower frequency is amplified and demodulated. Because most of the amplification is at this lower fixed frequency, there is much less likelihood of problems arising from feedback of this frequency to the input of the receiver, and both sensitivity and selectivity can be improved. The use of automatic gain control (AGC) reduces the effects of varying strength of received signals. FM receivers also use automatic frequency control (AFC) to maintain the correct oscillator frequency.

Stereo Radio

Stereo sound from discs and from tape was well established by the 1950s, but stereo sound broadcasting took a little longer. Stereo sound at its simplest means that two microphones are used for a recording, each picking up a slightly different signal. When these two signals are used to power separate loudspeakers, your ear detects a subtle difference in the sound; the effect on the ears is like that of 3D cinema on the eyes. Stereo recording on tape uses separate tracks for the signals; and on disc uses two tracks on one groove. We will look at these systems in Chapter 7.

Stereo broadcasting could be achieved by using two separate transmitters, one for each channel, but this would be very wasteful and the stereo effect can be disturbed if the two frequencies are subject to interference at different times or to different extents. Though the first experiments in stereo broadcasting in the UK (in the 1950s) used separate frequencies, it was obvious that stereo broadcasts would not succeed unless two conditions were met. One was that a single frequency must be used and the other was that users of mono receivers should still be able to pick up a mono signal when they tuned to a stereo broadcast. As it happened, these were very similar to the conditions that had been imposed on color television systems in the USA (see Chapter 8) in 1952, and similar methods could be adopted.

The system that evolved in the USA and which was adopted with only a few modifications in the UK was one that used a **sub-carrier**. As the name suggests, this is a carrier, which can be modulated with audio signals, and which is then itself used to modulate the main carrier. The frequency of the sub-carrier must be somewhere between the frequency of the main carrier and the highest frequency of the signals. The other important step was how to use the left (L) and right (R) channel information. The solution was to modulate the main carrier with the sum of the channel signals, L + R, which allowed any receiver tuned to the correct frequency to pick up this signal, which is a mono signal.

The stereo information was then contained in another signal which was the difference between the channels, L − R. This difference signal has a much smaller amplitude than the L + R signal and is needed only by stereo receivers, making it suitable for modulating on to a sub-carrier.

Figure 6.12 shows the frequency bandwidths of a stereo transmission before modulation on to the main carrier. The bandwidth that will be needed by the mono signal (L + R) is 15 kHz, and whatever other frequencies we use must not overlap this set. A single low-amplitude 19 kHz signal is also present; this is the **pilot tone** and it is used at the receiver (see later). The L − R

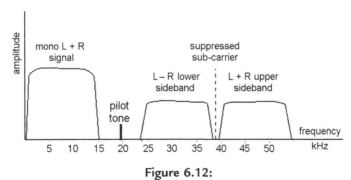

Figure 6.12:
The spectrum (graph of amplitude plotted against frequency) for a stereo FM radio signal. This shows the pilot tone and the sidebands of the sub-carrier which are added to the mono signal. This mixture of frequencies is modulated on to the main carrier at around 100 MHz

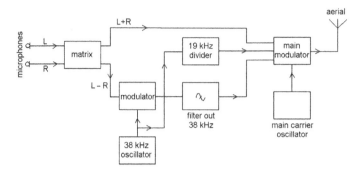

Figure 6.13:
A block diagram for the stereo FM transmitter. The L and R signals are mixed to produce the mono signal L + R and the smaller difference signal L − R which is modulated on to the sub-carrier

signals are **amplitude modulated** on to a 38 kHz sub-carrier, and this sub-carrier is then removed (making it a **suppressed carrier**), leaving only the sidebands of the modulated signal along with the pilot tone. These sidebands also need a bandwidth of about 15 kHz on each side of the 38 kHz mark. This set of different signals in different frequency ranges is used to frequency-modulate the main carrier of around 100 MHz.

Why remove the sub-carrier and supply a 19 kHz sinewave signal? The sub-carrier contributes nothing to the signal; it carries no information. It does, however, require transmitter power, and removing the sub-carrier makes an appreciable saving, allowing the transmitter to carry more of the useful sideband signals. The L − R signals cannot easily be demodulated, however, without supplying a 38 kHz signal in the correct phase, and this can be supplied locally at the receiver providing there is a small synchronizing signal available. This is the purpose of the 19 kHz pilot tone, whose frequency can be doubled in the receiver to 38 kHz and then used to correct the phase and frequency of the local oscillator in the receiver. The transmitter power used for the 19 kHz tone can be small, because the amplitude of this signal is small. As a bonus, the 19 kHz wave can be used at the receiver to turn on an indicator that shows that a stereo signal is being received.

Figure 6.13 shows the block diagram for transmission. The separate L and R channel signals are added and subtracted in a circuit called a matrix to give the L + R and L − R signals, and the L + R signal is used to frequency-modulate the main carrier. A master 38 kHz oscillator is used to provide the carrier for the L − R signals, and the 38 kHz signal is also used to provide a 19 kHz signal for the pilot tone, which is also frequency-modulated on to the main carrier. The modulated L − R signal is passed through a filter which removes the 38 kHz sub-carrier, so that the sidebands can also be frequency-modulated, along with the pilot tone, on to the main carrier. The modulated main carrier is then amplified and used to supply the transmitting antenna.

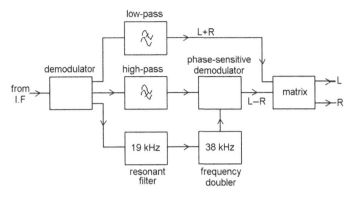

Figure 6.14:
A block diagram for the decoding portion of a stereo FM receiver. Only the portions from the FM demodulator are shown, as the remainder is a standard superhet block diagram

Figure 6.14 is the block diagram of the stages following the demodulator for a stereo FM receiver. The early stages of the receiver are exactly the same as they would be for a mono FM receiver, right up to the point where the FM signal is demodulated, providing three sets of signals. A low-pass filter separates off the L + R main signal and a resonant circuit separates out the 19 kHz pilot tone, and this is used to control the phase and frequency of a 38 kHz oscillator. Finally, a high-pass filter separates out the sidebands of the L − R signals.

A circuit called a **phase-sensitive demodulator** has inputs of the 38 kHz carrier frequency (obtained by doubling the 19 kHz pilot tone or by using it to synchronize an oscillator) and the L − R sidebands, and its output is the L − R signal itself. The L + R and L − R signals are fed into another matrix circuit, producing the L and R signals at the outputs. If you wonder how this is done, think of what happens when you add the inputs:

$$(L + R) + (L - R) = 2L$$

and also subtract them:

$$(L + R) - (L - R) = 2R$$

so providing the separate L and R signals.

Snags? The stereo signal is more easily upset by interference, particularly car ignition and other spark interference. A mono signal can use a low-pass filter to remove much of the effects, but this would remove the L − R signals from a stereo transmission. In addition, the noise level of a stereo transmission is always higher because the bandwidth is greater, and noise depends on bandwidth, the only noise-free signals are those with zero bandwidth, and signals with no bandwidth carry no information. Many FM receivers will automatically switch to mono reception when the noise level is high.

■ **Note**

Over the early years of the twenty-first century, FM is certain to be replaced by digital radio, mainly because governments all over Europe have sold the rights to the old FM carrier frequencies to mobile phone companies to provide an alternative to cables between mobile masts. A date in 2015 has been suggested for the UK, but the change will be slow because most digital receivers will sound no different. The uptake of digital radio has been patchy and politicians are reluctant to be labeled as the people who made us change to a system that is perceived as inferior for reception in many areas.

■

In fact, because the transmitters for digital radio in the UK use the old ITV transmitters (in the days before UHF television) the coverage is poor. This was not such a problem for television because houses had antennae on the roofs. Digital radio receivers that use a simple telescopic whip antenna receive a weak signal, and this can cause a most objectionable sound like boiling mud. In addition, digital radio transmitters often compress the sound signal (see later) so much that the quality suffers compared to a good FM receiver. Some misleading advertising suggests that digital provides a better quality, but this has certainly not been borne out by experience.

Digital radio was supposed to be at its best in cars, where its use eliminates the need to keep retuning to a local FM station as the car leaves one transmitter region and approaches another. The main advantage of digital radio is that a large number of digital radio transmitters can be used in a small amount of bandwidth without interference problems. This is more of an advantage for the broadcasters than for the listeners. Car manufacturers have been very reluctant to make the change, and have had to be bullied into it. The only real use for a digital car radio is that you can cruise around to find where the signal can be picked up, and car manufacturers will probably play safe and install radios that can be switched to FM when the digital signal is unobtainable or of low quality.

Internet Radio

Another possibility for digital radio broadcasting is the use of broadband Internet. This obviously requires a computer, and the sound from computer loudspeakers is fairly poor unless you have invested in high-quality units. With WiFi (Internet signals distributed from a small wireless transmitter in your house) the signals can be relayed to audio equipment, providing a signal quality that can be superior to the DAB signal. If you want to check out what can be achieved, look at the website http://www.shoutcast.com/, which is just one of the many sites offering radio by Internet. At the time of writing Internet radios (not part of a computer, but needing a WiFi signal) were just being advertised in the UK.

Summary

Stereo FM radio that is compatible with mono broadcasts is made possible by the use of a sub-carrier. The main signal consists of the sum of L and R channel audio signals, and this is a mono signal. The difference signal, L − R, is much smaller and is modulated on to a 38 kHz sub-carrier, and this carrier is filtered out, leaving only the sidebands. A 19 kHz pilot tone is added as a way of synchronizing a local oscillator at the receiver. All three signals are frequency-modulated on to the main carrier. At the receiver, the L + R signal can be used as a mono signal, and if stereo is required, the pilot tone has its frequency doubled and is used to synchronize the phase of a 38 kHz oscillator. This, along with the sidebands of the sub-carrier, can be used to demodulate the L − R signal, and combining the L + R and L − R signals provides separated L and R audio channel signals. The pilot tone can be used also to switch a stereo indicator lamp on or off. FM will eventually be replaced by digital radio (see Chapter 16).

CHAPTER 7

Disc and Tape Recording

▪ Note

Most of this chapter is now of historical interest because of the immense impact of digital methods on sound recording. Nevertheless, it is important to understand why the switch to digital was made, and this is possible only if you know what the problems of analog were (and still are) and how they have been dealt with. In addition, there is a vast number of analog recorders still around, and they shall quite certainly need repairs and maintenance for years to come.

▪

Early Gramophones

The first gramophones were totally mechanical. The sound that was being recorded was caught in a metal horn and used to vibrate a membrane at the narrow end of the horn. The vibrating membrane, in turn, moved a sharp stylus vertically on a revolving wax cylinder, driven by clockwork. This type of recording is known as 'hill and dale', and was also used on flat discs. To replay, the stylus was moved back to the start of the track, and the wax cylinder or disc was rotated again. This was the basis of the Edison Phonograph, and the wax cylinders that were cut for these machines are now treasured antiques.

Wax cylinders were fragile and difficult to mass produce, and the system was later applied to wax discs cut laterally so as to make a wavy groove, a system pioneered by Emile Berliner, who was producing these discs in 1892. The discs had the considerable advantage of being easier to store and pack than the cylinders and they also avoided the problem of the stylus jumping from one peak of a hill and dale recording to the next. The discs could easily be reproduced by electroplating and pressing to make shellac discs with the same imprint of a wave. Playing a record was simply the reverse of this process, spinning the shellac disc at a steady speed, eventually standardized as 78 revolutions per minute (r.p.m.), and picking up the sound using a needle in the groove whose movement vibrated a membrane at the end of a horn. Gramophones like this featured in many households in later Victorian times and were still in production, particularly in portable form, in the 1930s. Edison's original idea of 1877 and Berliner's improvements had a long working life.

Electronics Simplified. DOI: 10.1016/B978-0-08-097063-9.10007-X

The use of tubes for amplification promised to solve one of the problems of the acoustic gramophone, which was the lack of any effective volume control. You could always reduce the volume by stuffing a few socks down the horn, but increasing it was out of the question, though a few inventors harnessed compressed air to make amplifying horns. The main focus for improvements, however, was the recording process. Trying to get a full orchestra around a brass horn was difficult, and no form of control was possible. The possibility of using more than one microphone, being able to control treble and bass response with filters and to ensure that the record tracks did not overlap (through overloading) spurred the development of electrical recording methods. By the early 1930s, virtually all records were being made using electrical methods and the systems that were developed also contributed to the addition of sound to silent films in the late 1920s.

Electrical Methods

Figure 7.1 shows a typical electrical recording system, and this block diagram is applicable right up to the time when compact discs (CDs) started to replace the older vinyl long-playing (LP) records. In the block diagram, several microphones are shown connected to a mixer, so that some control can be obtained, allowing the sound balance of a soloist and of different sections of a band or an orchestra to be achieved. This type of system also makes it possible to reduce background noise, like the coughs and sneezes of an audience, to some extent. The mixer stage is followed by amplification and then by filtering (of the lowest and highest notes) and compression stages (to avoid overcutting), then by a driver (a power amplifier). The signals from the driver operate the cutter head which, as the name suggests, cuts a track in the master disc, which can be metal, wax, or plastic. The power that is needed for a good-quality cutter can be very large, and in the latter days of LP discs cutter-heads might require up to 1 kW of power.

The compression stages need some explanation. A sound source such as an orchestra has an immense range of amplitude (typically 100 dB or more) from its softest to its loudest, and this range simply cannot be recorded on to a disc (or a tape). Taking the disc example, the softest

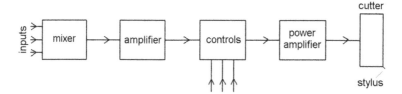

Figure 7.1:
Principles of disc recording. The signals from different microphones are mixed together, amplified and filtered (using potentiometer controls), and applied to a power amplifier that drives the cutter head. This cuts a master disc which is then used to produce other masters for pressing copies in vinyl plastic

sounds leave so little trace that only the noise of the needle on the disc can be heard on replay. The loudest sounds overdrive the cutter so that the waves on one part of the track overlap and break into an adjacent part of the track, making the record unplayable. The solution is to decrease the range, boosting the amplitude of the softest sounds and reducing the amplitude of the loudest, and this is the action called **compression**. In addition, even if the range of amplitudes is fairly small, the treble has to be boosted because most of the noise of a disc is in the form of a high-pitched hiss which would otherwise drown out the highest notes of music, and the bass has to be reduced because it is more likely to cause track overlapping.

The record player then has to reverse these processes, boosting the bass frequencies (typically in the range 30–120 Hz) and reducing the treble (typically in the range 4–10 kHz), and this process, called **equalizing** or **equalization**, is possible only if electronic methods are used; you cannot use old socks (or even new ones) for this task. In addition to equalizing, variable controls for treble and bass are needed to adjust the sound to compensate for the size, shape, and furnishing of the room in which it is heard. Figure 7.2 shows a block diagram for a record player of good quality with separate bass and treble controls. The signal is obtained from a pickup, another type of transducer which is similar in construction to a microphone, but using the vibrations from a stylus on the record groove rather than from a membrane.

Any pickup that is of reasonable quality suffers from the disadvantage that its output is a very small signal, in the region of a few millivolts at most, so that a fair amount of amplification is needed. The equalization, volume, bass and treble controls have to be applied at a stage where the amplitude of the signals is reasonably large, a volt or so, and the final portion is power amplification to drive a loudspeaker. This block diagram is, once again, one that has remained much the same from the 1930s to the 1980s, because the change from the older shellac 78 r.p.m. record to the $33^{1/3}$ r.p.m. long-playing vinyl disc required no change in the basic methods for a single channel (mono) signal. As for radio, changing first from tubes to transistors and then to integrated circuits (ICs) made no difference to what was being done, only to the details of how it was achieved.

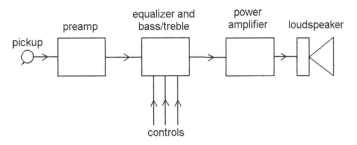

Figure 7.2:
Disc replay. The feeble signals (often less than 1 mV amplitude) from the pickup are amplified, and then equalized. At this stage, tone and volume controls can be used. The power amplifier then provides the drive for the loudspeaker

Summary

The system of disc recording pioneered by Berliner used a groove cut in a shellac (or later, vinyl) disc and modulated by the sound waves. For good-quality recording, the bass amplitude must be reduced and the treble frequencies boosted on recording and the opposite actions carried out on replay. Electrical recording and reproduction offers much more control over the process and has been used almost universally since the 1930s.

Hi-Fi and Stereo

In the late 1940s the word 'hi-fi', an abbreviation of 'high fidelity', burst in from the USA, and brought new life to the old gramophone. The spirit of hi-fi was that the sound which you heard at home should be as good as it was in the studio or concert hall. The ideal of perfection was always unattainable, but the principle was good: the quality of most gramophones at that time was abysmal, as bad as the quality we now put up from small transistor radios and from loudspeakers in almost every shop (with a few honorable exceptions). The nearest anyone ever got to reasonable quality at that time was in some cinemas (when they played music from a disc before or after the film) and better sound was a feature of the original version of Disney's *Fantasia* (1940), which could at that time be shown in full glory only in the very few cinemas that were equipped for stereo sound.

Hi-fi in these days required large loudspeakers (**woofers**) that could faithfully reproduce the bass notes, along with small ones (**tweeters**) to do justice to the higher notes. It also required good pickups, good preamplifiers, separate treble and bass controls and, above all, an excellent power amplifier stage. All of these requirements could be satisfied by good designs using tubes, and the amplifier that set the standard in the 1950s for the UK was the Leak, with its (then) remarkable figure of less than 0.1% distortion. In these days, home construction was popular, and many firms that became well known started with a home-made system which then was put into quantity production. Some of the amplifiers of that time are now valuable collectors' items, even the kit models that were sold for home assembly.

Hi-fi was about sound quality, and quality cannot be indicated in a block diagram, so that what we have seen in Figure 7.2 still holds. The development of stereo sound, however, made some changes. The principle was not new, and the idea of supplying each ear with a slightly different sound pattern had been established as long ago as the start of the twentieth century. Stereo creates an impression that the sound is no longer coming from one small space, and all the first listeners to stereo talked about a 'feeling of space', a 'broad band of music', and so on. What was lacking in the early days, however, was any way of achieving stereo sound on a standard gramophone disc, though the principle that was eventually adopted was based on one patented in 1936 by A.D. Blumlein (whose list of patents covers almost everything we think of as modern electronics).

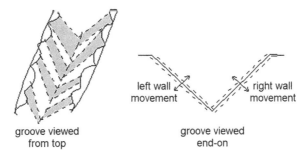

left wall movement

right wall movement

groove viewed from top

groove viewed end-on

Figure 7.3:
Stereo disc recording principles. The walls of the groove are at 90° to each other, and they are separately modulated with L or R signals The principle is that two mechanical motions at 90° to each other do not interfere with each other, so a pickup can detect L and R signals that are quite separate

Disc stereo recording uses a cutting head whose stylus can be vibrated in two directions that are at right angles to each other, cutting tracks in the walls of the groove. With the signals from two separate (left and right) microphones used to control the currents, one wall of the groove contains a left track and the other a right track (Figure 7.3). Replay also uses a single stylus that is connected to two pickup elements, each sensing movement in one of two directions at right angles, so that the outputs are of left and right signals. A movement along one axis has no effect on the axis that is at right angles to it, so that there should be no interference between the signals.

The stereo gramophone block diagram looks as in Figure 7.4, with separate L and R signals from the pickup passing through their own independent amplifier chains. Each channel (L or R) has treble and bass controls, and these are normally ganged (meaning that they are

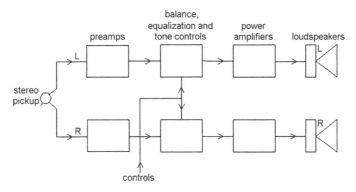

preamps

balance, equalization and tone controls

power amplifiers

loudspeakers

stereo pickup

L

R

L

R

controls

Figure 7.4:
A block diagram for a stereo disc replay system. This consist of two separate amplifiers, sharing a power supply, and with volume and tone controls ganged so as to operate together. There must also be a balance control to allow for loudspeaker positioning and other factors that may bias the sound to one side or the other

potentiometers that use the same spindle so that they rotate together). The volume controls can be separate, but it is more usual to gang them also and use an additional **balance** control to set the ratio of left to right volume. The balance control is normally set so that the main sound appears to come from between the two loudspeakers that are arranged one to the left and one to the right of the listener(s).

The coming of stereo caused a momentary halt in the hi-fi process, because listeners were more impressed by stereo of any kind than by high-quality single-channel sound (and this has not changed appreciably over the years). Another factor that has limited the appeal of high-quality sound has been the emergence of pundits who have brought to sound reproduction the flowery words and the mystique of wine tasting (but often with little justification). The plain truth is that it is easier now to achieve good-quality sound than it has ever been in the past, and without elaborate and costly equipment. Achieving the best possible quality is, as always, quite another matter, and if money is no object you can obtain really superb reproduction; but the outcome is often quite utterly disproportional to the cost, as you may have to redesign (or construct) the room in which you listen to cut out unwanted echoes and other disturbances.

Summary

The popularity of hi-fi as a hobby in the late 1940s led to an interest in better standards of reproduction, boosted by the vinyl LP record and the use of stereo recording. Hi-fi required good pickups, well-manufactured records, good amplifiers, and good loudspeakers, and though the transition from tubes to transistors (and to ICs) was slower than it was for other branches of electronics, it eventually happened. Vacuum tube amplifiers are, however, still available because some listeners prefer a distorted sound from a tube amplifier to a less distorted sound from a modern IC and transistor amplifier.

Tape and Cassette

We often think of magnetic tape recording as comparatively new, but it was invented by the Danish engineer Valdemar Poulsen (1869–1942) in 1889, and a (wire) recording that was made in 1900 by Emperor Franz Josef of Austria–Hungary has survived. The principle (Figure 7.5) is simple enough. A ring of magnetic material, with a coil of wire wound round it, will be magnetized when current flows in the coil. If there is a small gap in the rim, then the magnetization around this gap will be very strong, and when the current through the coil is alternating current (AC), the magnetization around the gap will also alternate in direction and strength. This concentrated magnetic field will affect any magnetic material near it.

Sound is converted, using a microphone, into electrical signals, and these are connected to the coil. A magnetic material, which originally was a steel wire, is drawn past the magnetic core at a steady speed, so that each portion of the wire will be magnetized to a different extent, depending on how much current was flowing in the coils when that portion of wire was drawn

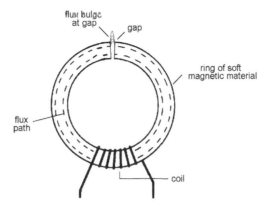

Figure 7.5:
A primitive tape-head in detail. The alternating current through the coil causes a strong alternating magnetic field at the gap which will magnetize any magnetic material that is in contact with the gap

past the core. To replay this, the process is reversed, connecting the coil to headphones. As the wire is drawn past the core, the changing magnetism causes currents to be created in the coil and these are heard as sounds in the headphones.

Poulsen's Telegraphone, as he called it, worked and was easier to use than a disc recorder with less risk of damage to a recording (unless you waved a magnet near it), but it never became a home item, unlike Edison's Phonograph (which used wax cylinders) and the later Berliner (flat disc) machines. The ability to record was not something that was important for entertainment in the days before radio, but the main point that made the wire recorder unattractive for the home was the need to use headphones: only one person could hear the faint recorded sounds, whereas the sound from a disc recording could 'fill a room with the full, rich notes' from its horn (the description dates from 1897). Poulsen's invention was used to a limited extent as an office dictating machine up to the 1920s but it was superseded for a considerable time by disc-based recorders.

The principle never totally died out, however, and radio companies experimented with wire recorders as a way of preserving important performances. The recording companies also took an interest, because recordings on wire could be edited, unlike recordings made on wax discs, and by using enough wire, a recording lasting twenty minutes or more could be captured. This was a significant advantage, because a wax disc could hold only about five minutes of recording. The limitations caused by the use of wire, however, were difficult to deal with, and these old recorders used (literally) miles of wire moving at high speed across the magnet core or **recording head**.

The rebirth of magnetic recording occurred in Germany during the 1939–1945 war. The BASF company, famous for aniline dyes, developed plastic tape with a magnetic iron-oxide coating that was immensely superior to iron or steel wire for recording uses, and could be used

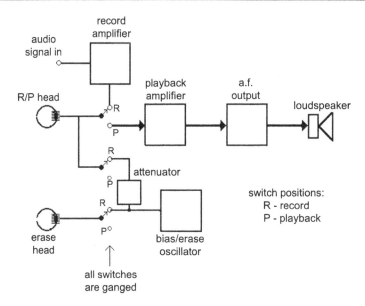

Figure 7.6:
A block diagram for a tape recorder, showing the record/replay switching. As is normal, the same head is used for both recording and replay, though better results can be obtained by using three separate heads (record, erase, and replay)

at much lower speeds of around 15 inches per second. Tape recorders of advanced design were found when the Allies invaded Germany, and were passed to electronics companies in the UK and USA for inspection. As a result, tape recorders became commercially available in the 1950s. In this respect, electronics firms were more commercially aware than the British motor industry, which was offered free manufacturing rights on the Volkswagen Beetle, but rejected it on the grounds that such a curious vehicle could not possibly be a commercial proposition. The outcome of that decision has been that VW now owns a large chunk of the former British motor industry, including Bentley (and BMW owns Rolls-Royce).

The record companies and radio companies were delighted with these developments in recording, and they started a demand, which still exists, for high-quality tape equipment. Apart from the editing convenience, the use of tape allowed for stereo recording (by using two tracks recorded on the tape) at a time when this was very difficult to achieve on disc, and also for sound effects that had not been possible earlier (produced, for example, in the BBC Radiophonics workshop, famous for its 'Dr. Who' theme). The use of tape recorders at this level fed down and resulted in an interest in sound recording at home.

Figure 7.6 shows a typical block diagram for a domestic tape recorder. The tape-head is a refined version of the coil and core arrangement, using a tiny gap in a magnetic metal to form a concentrated magnetic field across which the tape is moved during recording. The same head is used for replaying, and a switch allows the head to be connected to the record or replay

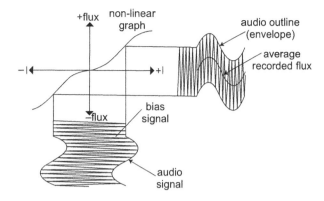

Figure 7.7:
The effect of bias with a high-frequency signal is to overcome to some extent the non-linear shape of the graph of magnetization plotted against signal strength

circuits. On professional tape recorders separate heads are used for writing and for reading. These blocks contain familiar portions, but the bias/erase oscillator block requires some explanation.

One important problem for magnetic recording is that magnetic tape is not a linear medium: a graph of recording current through the coil of the recording head plotted against the stored magnetism is not a straight line, more like an 'S' shape, with no retained magnetism at all for small currents (Figure 7.7). This makes the replayed sound appear impossibly distorted, even for speech. Early on, this had been tackled by adding a permanent magnet near the head, which improved matters but not to such an extent that tape could be considered good enough for recording music. The problem was solved by adding a **bias wave** at a high frequency (too high for the human ear to detect) of around 50–100 kHz. When this is done, using an amplitude that has to be carefully set for the type of tape and the tape speed, the results can be acceptable. With meticulous adjustment and good circuit design, tape recording could be good enough for making the master recordings for gramophone records, and was used in this way from the 1940s onwards. Making the quality acceptable on home recorders which had to be produced at a reasonable price was another matter.

■ Note

The upper limit of the range of frequencies that can be recorded is affected by the tape speed and the head gap. Achieving a good response for high frequencies demands either a very small head gap (a gap of 0.001 inches is an enormous gap!) or fast tape speeds. The other problem is noise, because the nature of magnetism makes tape a noisy medium. Noise is less of a worry on wide tape, but the trend has been to use narrower tape, often with four tracks so that the track widths are very narrow.

As it happened, the novelty for home recording on tape wore off, mainly because the tape, contained on open reels, had to be threaded into the recorder, and because there was little demand for recordings on tape to match the range and variety of gramophone discs. A whole generation of low-cost open-reel tape recorders was eventually scrapped when Philips invented and marketed the compact cassette in 1961. The principle is the same and so the block diagram is the same, but what has improved is convenience.

Cassettes originally offered low-quality sound, using narrow tape used at a low speed, and the original intention was to use the machines as dictaphones. The convenience of the cassette, however, led to recorders being marketed both for home recording and for replaying prerecorded cassettes, and in the following twenty years remarkable strides were made in improving tape, methods, and mechanisms until it was possible in the 1980s to claim with justification that cassette tapes could be used as part of a hi-fi system. The introduction of digital audio tape (DAT) made no impression on compact cassette sales, and all recordings that were available on other media (vinyl disc or CD) were at that time also available on compact cassette. A miniature form of recordable CD, called the Minidisc (MD), was available at one time but has now passed into history.

Summary

Tape and cassette recording of sound is not new, but the methods that are used to make the sound of acceptable quality are of recent origin. All magnetic recording makes use of a magnetic material being moved past a recording/replay head which consists of a metal core with a narrow gap and a coil of wire, and the construction of this head is very exacting if good-quality recording is required. Though open-reel tape recording is obsolete for domestic use, cassettes became a standardized and accepted way of distributing recorded music and their use has not completely vanished even now that CD and MP3 are the standard audio formats.

Cinema Sound

In the twentieth century, no electronic system before television made such an impression as cinema sound, films with a sound track, in 1928. The successful system, called Phonofilm, was invented by Lee De Forest, who had also invented the first amplifying radio vacuum tube, and the principles did not change until the use of magnetic tape tracks in more recent times. Early systems had suffered from inadequate volume and from synchronization problems, but De Forest solved both problems by using the film itself to record the sound and by making use of tubes for amplification.

The block diagrams for a cinema sound system of the traditional type (now almost obsolete) are illustrated in Figure 7.8. On recording, the sounds are converted into electrical signals using microphones, and mixing is carried out as required. The amplified signal is used to drive

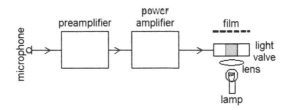

Figure 7.8:
Recording cinema sound, using the film to carry the sound track. This depends critically on the design of the light-valve

a 'light-valve', which opens or closes according to the amplitude of the signal wave at its input. These light-valves were originally electromagnetic, though other principles have been used. By placing the light-valve near the edge of the film, illuminated by a light beam, a strip of otherwise unused film can be exposed to light that is controlled by the light-valve, creating a wave pattern on this strip of film, the sound track.

Since these early days, two separate principles have been used. One used the light-valve to vary the **width** of the light beam (variable-width system) that reached the film, while the other used the valve to control the **brightness** of the light (variable-density system). Whichever method was used on recording, the developed and printed film contained a strip of sound track (Figure 7.9) situated between the frame edge and the sprocket holes. This strip of film has been exposed so that it contains a picture of the sound in the form of variations of either width or of blackness (Figure 7.10).

This sound track can be played back using a **photocell**, a type of transducer that responds to light either by generating a voltage or by allowing current to pass from a voltage supply. On playback, the light of the cinema projector hits the sound track as well as the main frames of the film, so that as the film moves the amount of light passing through the sound track to the

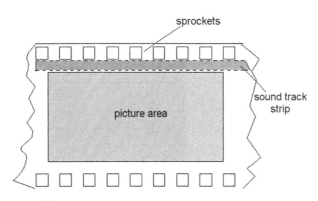

Figure 7.9:
Location of the sound track on a 35 mm film

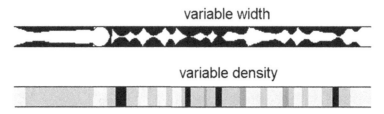

variable width

variable density

Figure 7.10:
Variable width and variable density sound tracks compared. The replay system will cope with
either type of track

photocell varies, generating a variable voltage or current which is an electrical signal that can
be amplified (Figure 7.11). The same playback system can be used whether the film uses
variable-width or variable-density sound recording.

Incidentally, the sound track that corresponds to any particular frame of a file is not placed next
to that frame. This is because film in a projector or camera does not run smoothly but in jumps.
Each frame is held for about 1/24 second for exposure and then moved rapidly, and this would
put a loud 24 Hz hum on to the sound track if we read the sound track at the point of projection
of the frame. Instead, the sound track is placed about 30 frames ahead of the picture at a point
where the film is moving steadily rather than in jumps. This solves the hum problem, but it
makes editing more complicated than it was for silent films, and it also causes problems if
a film breaks and has to be repaired by taking some frames out, because the sound track that is
removed does not then match the removed picture.

Magnetic recording, using a magnetic stripe on the edge of the film (preceded by a system in
which a reel of magnetic tape was wound alongside the film), was used between 1952 and

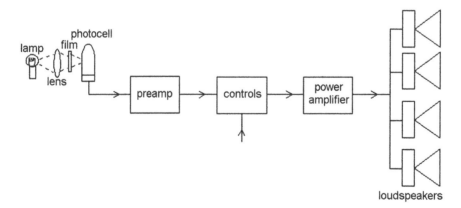

Figure 7.11:
The replay arrangement uses a photocell whose (feeble) output signal is amplified. Cinema sound
requires large amplifiers to provide power to a bank of loudspeakers spaced around the cinema

1999. This was eventually abandoned because the magnetic stripe was vulnerable to wear and tear to a much greater extent than an optical track. At the time of writing several types of digital recording exist using optical recording and most films are released with several types of optical tracks and the older analog optical track as well. Digital tracks are recorded on the side of the film opposite the analog tracks, and also in the spaces between sprocket holes. Several versions of optical digital recording are due to the well-known recording specialist, the Dolby Laboratories.

Home Cinema

The availability of large-screen television receivers has triggered a market for home cinema for watching high-definition television or Blu-ray recordings (see later) along with multichannel sound. The familiar part of the equipment has to be a large-screen television receiver, usually a plasma type rather than liquid crystal display (LCD), and set up for high definition (HD) as well as standard definition. A Blu-ray disc player (which can also play the older type of DVD) forms an essential part of the video package. For sound, a 'surround-sound' system is needed using at least five loudspeakers (front left and right, rear left and right, and a sub-woofer, for very loud low-pitch unmusical sounds such as explosions, which can be placed centrally). A controller/amplifier is also needed for this setup so that the audio signals from the television signal or Blu-ray player can be allocated to the correct loudspeakers.

The results, like the prices, can very enormously, but even the low-cost systems can provide an enjoyable way of watching movie films at home. As for all sound systems, however, the domestic arrangements (furniture, curtains, carpets) can have a considerable effect on the sound, so that real enthusiasts will set aside a room as a cinema room, with acoustics that enhance the experience. It is possible to spend as much as it might cost you to visit your local commercial cinema nightly for several years.

Noise

Noise is one of the persistent problems for any type of recorded sound, as we have indicated already. A really old recording on shellac disc will consist so much of hissing and scraping that you wonder how anyone could listen to it with any enjoyment, and though the noise level of discs had been reduced by the 1940s it was the main factor in the popularity of the long-playing vinyl disc which replaced the older type. The lower speed of the LP, along with the use of a long-life diamond or sapphire stylus and equalization circuits, kept the noise level of the LP reasonably low, and this seemed satisfactory until we heard (or didn't hear) the noise level of CDs.

The noise level on early cassette tapes almost rivaled that of the old discs. Tape noise is always a problem, and it was dealt with to some extent on professional equipment by using wide tape moving at high speed. The width of a cassette tape allocated for four tracks (two in each direction) is only 1/8 inch, whereas an open-reel machine will use at worst two tracks on 1/4 inch tape, or use much wider tape such as the 1/2 inch type (now used for video) or more. In addition, the frequency range that can be recorded on tape is very limited when the tape speed is low. Cassette tape moves slowly, at about 1 7/8 inches per second (i.p.s.) rather than the 15 i.p.s. or 30 i.p.s. of professional tape equipment, and this can restrict the highest frequencies that can be replayed to as low as 6 kHz or less. This inability to record or replay the higher frequencies is also controlled by the width of the gap in the tape-head, and a large gap means one-thousandth of an inch. Tape-head gaps are usually stated in micrometers, with 1 μm being a thousandth of a millimeter.

Noise Reduction

Development of better tape materials has greatly improved the noise performance of tape, and we can now make tape-heads with gaps that are in the region of $1-10$ μm. Even with all that, however, tape could not be considered suitable for serious use without noise-reduction circuitry. Several schemes have been used in the past, but the only ones that have survived have used equalization methods, deliberately altering the signals before recording and reversing the action on replay. The best known of these noise-reduction systems is Dolby, named after the British engineer Ray Dolby (who at one time also worked in the development of video recording at Ampex Inc.). The Dolby B noise-reduction system is almost universally used for prerecorded cassettes, and other versions such as C or HX Pro are used on more expensive equipment. There was also another system, **dbx**, which was astonishingly effective, but had some drawbacks that prevented commercial success.

Figure 7.12 illustrates the Dolby principle. Tape noise is at a low amplitude and it is concentrated in the higher frequencies. The amplitude of the noise, however, is almost the same as the amplitude of the softest music, and the hiss that you hear when you turn up the volume to hear better is very noticeable because the human ear is particularly sensitive to this range of frequencies. In addition to the noise problem, tape will overload if too large an amplitude of signal is recorded.

On recording, the Dolby circuits split the electrical signal into separate channels. The lower frequency signals pass unchanged. The low-amplitude signals in the higher frequency range are boosted, making their amplitude greater than they were in the original signal. On the C and HX Pro systems, signals that have an amplitude that would cause overloading are reduced, so the result is a signal whose smallest amplitudes are still above the noise level and whose maximum amplitude does not overload the tape. When the Dolby recorded tape is replayed, the processing that has been carried out on recording is reversed.

The attenuation of the signals that have been boosted will also reduce the noise signals from the tape so that they are almost undetectable and if the more advanced systems have been used, the

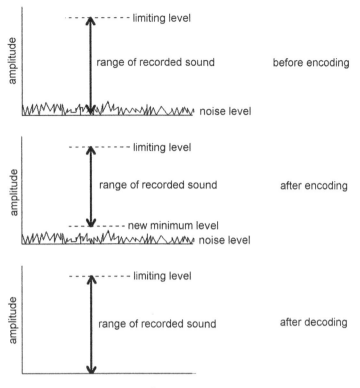

Figure 7.12:
Noise reduction on cassette recorders. The sound signals are processed so as to reduce the range, in particular to ensure that no signals fall to a level as low as that of the noise. When the signal together with noise is expanded, the noise almost disappears

tape will also deliver sound that has a much greater range (from softest to loudest) than is possible when no processing has been used. Though the C and HX Pro systems offer better noise reduction, the Dolby B system is almost universal on prerecorded cassettes because it does not alter the sound too much. This means that if you replay a Dolby B recording on a player that is not equipped with Dolby decoding the sound does not sound distorted, just a little biased to treble. This is not something that could be said for the other systems (though undecoded dbx can be acceptable in a noisy car because the volume is almost constant).

■ Note

These noise-reduction systems are often referred to as **companders**, because the amplitude range is compressed when the recording is made and expanded again when the sound is replayed.

■

Because of Dolby, cassette tapes can be used as part of a serious hi-fi system, and cassette tapes that have been recorded using systems such as Dolby C or HX Pro can almost rival a CD for low noise, though not for sound amplitude range. Dolby noise reduction is also used extensively in cinema sound, nowadays as a digital–optical system.

■ Note

Noise and noise-reduction systems have now become much less important with the adoption of digital methods (such as CD) for sound recording other than for cinema.

■

Summary

Noise is the enemy of all recording and broadcasting systems. Tape noise is particularly objectionable because it consists of a hiss, a type of sound to which human ears are particularly sensitive. For professional recording, the use of wide tape overcomes the problem to a considerable extent because a wider tape has a larger magnetized area and so produces a larger signal that 'swamps' the noise. The narrow tape used on cassettes, however, has a high noise level that cannot be overcome completely by using better tape materials. Noise-reduction systems operate on the principle of selectively boosting the amplitude of the recorded sound when it is recorded and decreasing it on replay, and a particularly effective method of doing this concentrates the boosting and reducing actions on the frequency range that is most affected.

Video and Digital Recording

The recording of sound on tape presented difficulties enough, and at one time the recording of video signals with a bandwidth of up to 5.5 MHz, and of digital sound, would have appeared to be totally impossible. The main problem is the speed of the tape. For high-quality sound recording, a tape speed of 15 i.p.s. was once regarded as the absolute minimum that could be used for a bandwidth of 30 Hz to about 15 kHz. Improvements in tape and recording head technology made it possible to achieve this bandwidth with speeds of around 1 i.p.s., but there is still a large gap between this performance and what is required for video or for digital sound recording. This amounts to requiring a speed increase of some 300 times the speed required for audio recording. Early video recorders in the 1950s used tape speeds as high as 360 i.p.s. along with very large reels of tape.

Analog video recording, even now, does not cope with the full bandwidth of a video signal, and various methods of coding the signal are used to reduce the bandwidth that is required. In addition, the luminance (black and white) video signals are frequency-modulated on to a carrier, and the color signals that are already in this form have their carrier frequency shifted (see Chapter 8 for more details of luminance and color signals).

For domestic video recorders, the maximum bandwidth requirement can be decreased to about 3 MHz without making the picture quality unacceptable, but the main problem that had to be

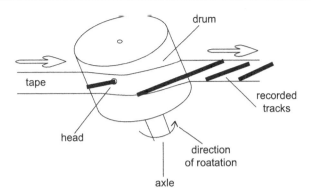

Figure 7.13:
Principle of rotary-head video recording. The two (or more) heads are mounted on a drum, and the tape is wrapped at a slight angle. This makes the head trace out a sloping track across the tape as the drum revolves and the tape is pulled around it at a slow rate

solved was how to achieve a tape speed that would accommodate even this reduced bandwidth. In fact, the frequency of the carrier ranges between 3.8 and 4.8 MHz as it is frequency-modulated to avoid the problems of uneven amplitude when such high frequencies are recorded on tape.

The brilliant solution evolved by Alexander Poniatoff (founder of the Ampex corporation) was to move the recording head across the tape rather than move the tape over a head. Two (now often four) heads are used, located on the surface of a revolving drum, and the tape is wound round this drum so that the heads follow a slanting path (a **helical scan**) from one edge of the tape to the other (Figure 7.13). The signal is switched from head to head so that it is always applied to the head that is in contact with the tape. At a drum rotation speed of around 1500 rotations per second, this is equivalent to moving the tape past a head at about 5 meters per second.

Though the way that the head and the tape are moved is very different from that used for the older tape recorders, the principles of analog video recording remain unchanged. The block diagram for a video cassette recorder is very different from that of a sound recorder of the older type, but the differences are due to the signal processing that is needed on the video signals rather than to differences in recording principles.

■ Note

Videotape is, at the time of writing, almost obsolete (as attested by the huge stacks of videotapes in charity shops) and has been superseded by digital versatile disc (DVD), the digital system that uses the same principles as CD. This is particularly suited to digital television signals, and in the UK has been used mainly for players of unnecessarily expensive discs (it costs less to press out a DVD than to record a tape). DVD recorders with a recordable and reusable disc are readily available and of good performance, and another answer to the need for domestic recording and replay has been to use a computer hard drive (magnetic disc) along with conversion circuits so that ordinary analog television signals (as well as digital signals) can be recorded digitally.

These hard drive units have typically a capacity of up to 45 hours, so that a single unit can cope with most domestic recording needs. With the switch to digital television in the UK complete in 2011 the use of hard-drive video recorders will be almost universal. A particular advantage of these hard-drive recorders is that (using buffer stages) they can record and replay simultaneously, so that when the unit is switched on it is possible to view a live program, place on hold while answering a telephone call or having a meal, and resume viewing later. The facility is available also for digital radios in the areas where reception is possible. Ideally, we might have a hard-drive recorder along with a DVD recorder so that really useful recordings could be preserved.

Later sound recorders (before the extensive use of CD recorders) for very high-quality applications used **digital audio tape** (DAT). This operated by converting the sound into digital codes (see Chapter 9) and recording these (wide-band) signals on to tape using a helical scan such as is used on video recorders. The main problem connected with DAT is that the recordings are too perfect. On earlier equipment, successive recording (making a copy or a copy of a copy) results in noticeable degradation of the sound quality, but such copying with DAT equipment causes no detectable degradation even after hundreds of successive copies. This would make it easy to copy and distribute music taken from CDs, and the record manufacturers succeeded in preventing this misuse of DAT (though not in the Far East). DAT recorders that were sold in the UK were therefore fitted with circuits that limited the number of copies that could be made, and the DAT system disappeared when recordable CDs and, later, DVDs were developed.

Summary

Tape as a recording medium hardly seems adequate for sound recording, and its use for video and for digital sound has been a triumph of technical development. As so often happens, however, the relentless progress of technology has made tape-based systems obsolescent just as they seemed to have reached their pinnacle of perfection.

CHAPTER 8

Elements of Television

Television

At the time of writing, the changeover to digital television in the UK is well under way and is due to be completed in the summer of 2011, and the changeover in Australia that started in 2001 should be complete by 2013. The change was completed in the USA in 2009.

The methods of coding the picture information for digital transmission are very different from those used in the older (analog) system, but many of the basic principles of television are much the same. In this chapter we will concentrate first on the older analog system because these receivers can have a long life and may still be in use (using a digital set-top box converter) over the lifetime of this book, and look later at the differences brought about by digital television.

■ Note

The main benefits of digital television are that it uses less transmitter bandwidth (allowing the government to sell off unwanted frequencies and to license new channels) and is more compatible with flat screens, computers, digital versatile discs (DVDs), and other newer technologies. Older receivers can still be used along with a set-top box, so that owners of satellite dishes had a head start in the use of digital before terrestrial systems (using the existing antenna) such as the UK Freeview system or US OTA became available. For most viewers, digital offers a noticeable improvement in picture quality, and is the only way of delivering high-definition (HD) pictures.

■

Scanning

Oddly enough, television has had a rather longer history than radio, and if you want to know more of its origins there was a book called *Birth of the Box*, now out of print, which dealt with the fascinating development of this subject. Almost as soon as the electric telegraph (invented in 1837) allowed messages in Morse code (invented in 1838) to be sent along wires, inventors tried to send picture signals, and this activity was spurred on by the invention of the telephone. Sound transmission along wires is simple, because it requires only that the sound waves be converted to electrical waves of the same frequency. A picture, however, is not in a usable wave

Electronics Simplified. DOI: 10.1016/B978-0-08-097063-9.10008-1

format, but by the middle of the nineteenth century the principle of **scanning** a picture had been established.

Definition

Scanning means breaking an image into small pieces, reading them in sequence, and coding each of them into an electrical signal.

Suppose, for example, that you drew a picture on a grid pattern such as in Figure 8.1. If each square on the grid can be either black or white, you could communicate this picture by voice signals to anyone with an identical grid by numbering each square and saying which squares were white and which were black. It may sound elementary, but this is the whole basis of both television and fax machines (which are almost identical in principle to the first television transmitters and receivers). This grid could be described as a 12×12 picture, but for television we need a much larger number of squares, typically about 500×300. Currently, computers use 640×480 as a minimum, with 800×600, 1024×768, or higher, more often used. Analog television more often specifies only the **vertical resolution** in terms of the number of lines (currently 625 in the UK and Europe, 525 in the USA), though not all of the lines carry picture information. This makes it difficult to compare the performance of analog television receivers

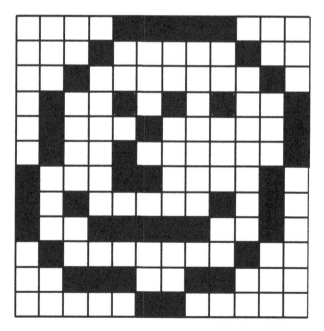

Figure 8.1:
A simple image consisting of black or white squares. We can imagine this being built up line by line by scanning across and down and deciding whether a square should be filled or not

with computer monitors. HD television displays can use 1024×600 for 16:9 pictures (see later).

Early television systems, patented around 1870 by Nipkow and Rosing in St Petersburg, and subsequently constructed by John Logie Baird in the late 1920s and early 1930s, used mechanical scanning (with rotating discs or mirrors) and were confined initially to still pictures of around 30 scanned lines, in black and white, with no shades of gray. Television as we know it had to wait until electronic components had been developed, and in particular, the receiver cathode-ray tube (CRT) and the camera tube.

By the early 1930s, however, Philo Farnsworth in San Francisco had demonstrated that electronic television was possible (at almost the same time as Baird had made a sound broadcast declaring that CRTs could never be useful for television). Also at that time, a form of radar had been used to measure the distance from Earth of the reflecting layers of charged particles in the sky (the Heaviside and Appleton layers). These two forms of technology were to play decisive roles in the development of electronics in the twentieth century.

The attraction of the CRT is that it can carry out scanning by moving an electron beam from side to side and also from top to bottom, tracing out a set of lines that is called a **raster**. Nothing mechanical can produce such rapid changes of movement as we can obtain by using an electron beam, so that most of the television pioneers realized that this was the only method that would be acceptable for reasonable definition, though the cinema industry of those days (which backed Baird) still believed that mechanical systems might prevail. By the 1930s, it was becoming acknowledged that any acceptable form of television receiver would certainly use a CRT, and the main research effort then went into devising a form of CRT that could be used for a television camera. The principles of such a tube were:

- a light-sensitive surface on which an image could be projected using a lens, causing electrons to leave the surface
- a scanning electron beam which could replace the electrons lost from the light-sensitive surface
- some method of detecting how many electrons were being replaced for each portion of the image (this is the signal current).

Without going into details, then, a camera CRT must allow the image projected on to the face of the tube to be converted into electron charges, and these charges are scanned by an electron beam from which it must produce a signal current or voltage that can be amplified so as to represent the brightness of each part of the scanned picture.

Synchronization

The problems of developing CRTs were only part of the story. A remarkable lecture in 1911 by J.J. Campbell Swinton outlined a television system that we can recognize today and which was

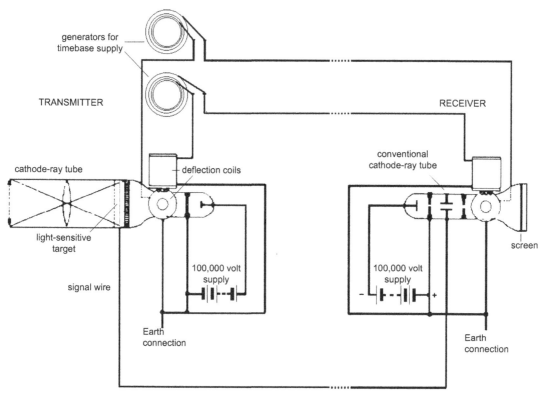

Figure 8.2:
A drawing of Campbell Swinton's proposal for an electronic television system

at least twenty years ahead of its time. It imagined that both camera and receiver would use CRTs, and for the first time emphasized that the scanning had to be synchronized, and that this could be done using start-of-scan signals that had also to be transmitted so that the scanning at the receiver would remain in step with the scanning at the transmitter. Figure 8.2 shows Campbell Swinton's drawing. If we can ever credit anybody with inventing television as we know it, the honors have to be shared between Campbell Swinton and Philo Farnsworth. This leaves Baird with the honor of inventing the word 'television' and showing that the 1870 system could be used as a primitive form of fax machine. Baird could rightly claim that he proved that television was possible.

■ Note

The use of camera tubes declined rapidly towards the end of the 1990s because of the development of light-sensitive semiconductor panels of the charge-coupled device (CCD) type. These have superseded the types of CRT used for television cameras, making cameras much lighter and less bulky. Their development has also made possible the camcorder that combines a television camera and video recorder in one package, and the most recent digital camcorders that fit into the palm of your hand. CCD units are also used in digital still cameras. For reception, flat-screen technologies have almost totally replaced CRTs.

■

We can make a block diagram (Figure 8.3) for a simple, more modern television system that follows closely along the lines laid down by Campbell Swinton. This shows a camera tube which is scanned by signals from a generator, and these signals are also sent to a transmitter along with the amplified vision (**video**) signals from the tube. These signals can be sent down a single wire (or they can be modulated on to a radio wave) to the receiver where the video signals are separated from the scanning signals, and the separate signals are applied to the receiver CRT. This is virtually a block diagram for the first electronic television systems used from 1936 onwards.

The most remarkable aspect of the development of electronic television in the 1930s was that so many new problems were tackled at the same time. The television signal needed a large bandwidth, thousands of times greater than anything that had been used in radio. The carrier frequency had to be high and very few radio tubes used at the time were suitable to be used at such high frequencies. Circuits had to be invented to generate the scanning waveforms, which needed to be of sawtooth shape (see Chapter 1). A system had to be worked out for allowing the receiver to generate its own sawtooth scanning waveform, but keep these perfectly synchronized with the scanning at the transmitter. Not until the space race would any team of engineers tackle so many new problems together on one project.

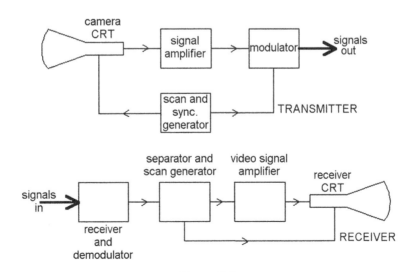

Figure 8.3:
A block diagram for a more modern version of a simple electronic television system

Summary

Though primitive television systems used mechanical scanning, television as we know it was impossible until electronic methods of scanning a picture were devised. The scheme outlined by Campbell Swinton in 1911 was the basis of modern television, and it made use of a form of CRT at both transmitter and receiver. This scheme also emphasizes the use of synchronizing signals that would be used by both transmitting and receiving equipment.

The Analog Television Waveform

The problems of analog television are illustrated by looking at the shape of the television waveform. Like any wave, this is a repeating pattern, but the shape is much more complicated than that of a sound wave. In particular, the wave contains some portions that have very sharp edges, and these edges are vital to synchronization. A typical single wave for a black-and-white television system of the 1950s is shown in Figure 8.4(a), along with a portion of the waveform that is transmitted at the end of a 'field' when all of a set of lines have been scanned and the beam is at the bottom of the tube face (Figure 8.4b).

The small rectangular portion of the wave in Figure 8.4(a) is the line-synchronizing pulse ('line sync'), and it is used in the receiver to ensure that the electron beam starts scanning across just as the first part of the video wave reaches the receiver. The video wave itself will have a shape that depends on the amounts and position of dark and light parts of the

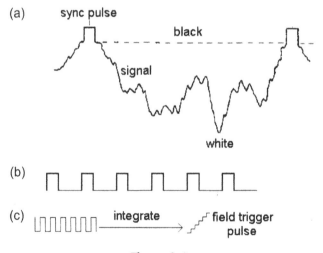

Figure 8.4:
(a) The standard analog television signal (for a black-and-white picture), showing the signal level for one line, and (b) the pulses that are transmitted at the end of a field. The signals are shown as they appear when modulated on to the carrier, with the tips of the sync pulses representing peak carrier amplitude. The effect of integrating the field pulses is illustrated in (c)

picture across one line. Mechanical television originally used a 30-line scan, but electronic television started in the UK with 405 lines (525 lines were used in the USA), a huge improvement in resolution that allowed relatively fine detail to be seen. The video amplitude ranges between black and peak white, and the synchronizing pulses are in the opposite direction, lower than the level that is used to represent black so that nothing can be seen on the screen when the pulses are transmitted.

At the end of a set of lines, the waveform changes. There is no video signal, just a set of more closely spaced synchronizing pulses. These are integrated at the receiver to generate a **field** synchronizing pulse which will bring the beam back to the top left-hand corner of the screen ready to scan another set of lines. As before, these pulses have an amplitude that is lower than black level ('blacker than black') so that they do not cause any visible disturbance on the screen.

As it happens, using a full set of 405 lines to make a picture was out of the question when television started in 1936 (and it would still cause problems on an analog system even now). The trouble is that a waveform consisting of 405 lines repeated 50 times per second needed too much bandwidth. Pioneers in the USA had even more to cope with, using 525 lines at 60 pictures per second. The answer that evolved on both sides of the Atlantic was **interlacing**. Interlacing, using the UK waveform as an illustration, consists of drawing the odd-numbered lines in $\frac{1}{50}$ second and the even-numbered lines of the same picture in the next $\frac{1}{50}$ second (Figure 8.5). This way, the whole picture is drawn in $\frac{1}{25}$ second, and the bandwidth that is needed is only half as much as would be needed without interlacing. Interlacing is still used on analog television and digital television pictures, though it is not used on computer monitors because on the small bright images of a monitor the use of interlace can cause a flicker which is visually disturbing.

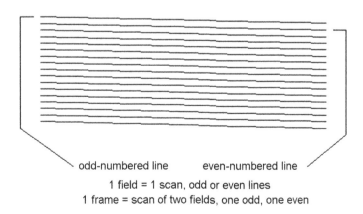

odd-numbered line   even-numbered line

1 field = 1 scan, odd or even lines
1 frame = scan of two fields, one odd, one even

Figure 8.5:
The principle of interlacing. This allows a more detailed analog picture to be transmitted without using an excessive bandwidth

■ **Note**

Digital television transmissions send out interlaced signals so that they can be displayed on television receivers that use a CRT. Modern flat-screen television receivers using liquid crystal display (LCD) or plasma displays do not need interlacing (which can cause problems) and so the received signal is deinterlaced. Computer monitors do not use interlacing at all. Eventually, interlacing will be discontinued for television broadcasts.

■

Summary

The analog television waveform is not a symmetrical wave of fixed shape. The synchronizing pulse for each line occurs at regular intervals, but the waveshape that follows this pulse depends on the distribution of light and shade in that line of the picture. The use of interlacing, scanning only half of the total number of lines in each vertical sweep, reduces the bandwidth by half without degrading the picture quality. Interlacing is not required for flat-screen displays, but is still used on the broadcast signal for compatibility with CRT receivers.

Transmission

We can look now at a block diagram of the studio and transmission side of black-and-white television in the days when CRTs ruled. We will look later at the additions that have to be made for color.

Television transmission at its simplest starts with a television camera, which is fed with synchronizing pulses from a master generator. Light from the scene that is to be transmitted is focused through a lens on to the face of the camera tube, and this light image is converted into an electron charge image inside the front section of the tube. The scanning electron beam discharges this, and the discharge current is the **video signal**. This is the starting point for the block diagram of Figure 8.6. The video signal, which is measured in microvolts rather than in millivolts, has to be amplified, and the synchronizing pulses are added. The generator that supplies the synchronizing pulses also supplies the scans to the camera tube. The complete or **composite** video signal that contains the synchronizing pulses is taken to the modulator where it is amplitude-modulated on to a high-frequency carrier wave.

Modern analog television systems modulate the signal so that peak white is represented by the *minimum* amplitude of carrier and the synchronizing pulses by the *maximum* amplitude. In the original UK television system the signals were modulated the other way round, with peak white represented by maximum carrier amplitude, but this method allowed interference, particularly from car-ignition systems, to show as white spots on the screen. When the system

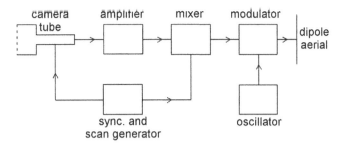

Figure 8.6:
Television transmission from camera to antenna. This block diagram shows a simplified version of the transmission of a monochrome picture from a camera of the older type using a camera tube

changed to 625 lines in 1968, maximum carrier amplitude was used for the sync pulse peaks in line with the methods used in other countries. Using this system, interference causes black spots that were much less noticeable, and the compulsory use of suppressor resistors in car-ignition systems has almost eliminated the main source of television interference.

The sound signal has been omitted from this to avoid complicating the diagram. The sound is frequency-modulated on to a separate carrier at a frequency higher than that of the vision carrier, and is transmitted from the same antenna.

Summary

The analog video signal is generated from a camera tube or a CCD panel and is amplified and combined with synchronizing signals to form the composite video signal. This is amplitude-modulated on to a carrier with the tips of the sync pulses represented by peak carrier amplitude. The sound signal is frequency modulated on to a separate carrier which is at a frequency 6 MHz higher than the vision signal.

Modern Television Displays

Cathode-Ray Tubes

The CRT is an active component that nowadays is just about the only device you are likely to come across that uses the same principles as the old-style radio tubes. The CRT converts an electrical signal into a light pattern, and though the principle was discovered in the 1890s the technology for mass production was not available until the 1930s and by that time was being pushed on by the needs of radar as much as by those of television. Nowadays, CRTs are still around in older television receivers, but they are rapidly being replaced by modern flat-screen displays that use semiconductors and are better suited to modern digital television signals.

Definition

The CRT can be used to convert variations of voltage into visible patterns, and is applied in instruments (oscilloscopes), for television, and for radar, though their use in these applications is rapidly dying out.

The three basic principles of the CRT are that:

- Electrons can be released into a vacuum from very hot metals.
- These electrons can be accelerated and their direction of movement controlled by using either a voltage between metal plates or a magnetic field from a coil that is carrying an electric current.
- A beam of electrons striking some materials such as zinc sulfide will cause the material (called a **phosphor**) to glow, giving a spot of light as wide as the beam.

■ Note

Do not be misled by the name: a phosphor does not contain phosphorus (though it can contain several nasty poisons). The name simply means a substance that glows.

■

Nothing is ever as simple as basic principles might suggest. There are very few metals that will not melt long before they are hot enough to emit electrons, and at first, only the metal tungsten was deemed suitable. Tungsten was used for radio tubes and for CRTs until a chance discovery that a mixture of calcium, barium, and strontium oxides, comparatively cheap and readily available materials, would emit electrons at a much lower temperature, a dull-red heat. By the early 1930s, 'dull-emitter' tubes were being mass produced, and some CRTs were being manufactured for research purposes.

Figure 8.7 shows a cross-section of a very simple CRT. The **cathode** is a tiny tube of metal, closed at one end and with that end coated with a material that emits electrons when it is red hot. A coil of insulated wire, the **heater**, is used to heat the cathode to its working temperature. Because the far end of the tube contains conducting material at a high voltage (several kilovolts), electrons will be attracted away from the cathode.

These electrons have to pass through a pinhole in a metal plate, the **control grid**. The movement of the electrons through this hole can be controlled by altering the voltage of the grid, and a typical voltage would be some 50 V **negative** compared to the cathode. At some value of negative grid voltage, the repelling effect of a negative voltage on electrons will be greater than the attraction of the large positive voltage at the far end of the tube, and no electrons will pass the grid: this is the condition we call **cut-off**.

Electrons that pass through the hole of the grid can be formed into a beam by using metal cylinders at a suitable voltage (Figure 8.8 shows a set of typical voltages for a small CRT).

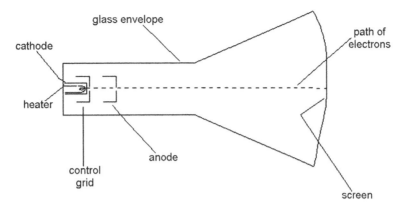

Figure 8.7:
A diagram of a very simple type of cathode-ray tube which can produce an electron beam that in turn will make a spot of light appear on the screen. Details of connections to a base have been omitted

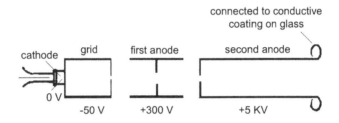

Figure 8.8:
Typical voltages on the electrodes of a small cathode-ray tube as would be used in oscilloscopes at one time

By adjusting the voltage on one of these cylinders, the **focus electrode**, the beam can be made to come to a small point at the far end of the tube. This end is the screen, and it is coated with the **phosphor** that will glow when it is struck by electrons. The phosphor is usually coated with a thin film of aluminum so that it can be connected to the final accelerating (**anode**) voltage. The whole tube is pumped to as good a vacuum as is possible; less than a millionth of the normal atmospheric pressure.

This arrangement will produce a point of light on the center of the screen, and any useful CRT must use some method of moving the beam of electrons. For small CRTs used in traditional oscilloscopes a set of four metal plates can be manufactured as part of the tube and these **deflection plates** will cause the beam to move if voltages are applied to them. The usual system is to arrange the plates at right angles, and use the plates in pairs (Figure 8.9), with one plate at a higher voltage and the other at a lower voltage compared to the voltage at the face of the tube. This system is called **electrostatic deflection**.

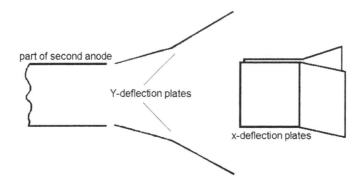

Figure 8.9:
Using metal plates to deflect the electron beam. The plates are sloped and bent to ensure that the deflected beam does not strike them

There is an alternative method for deflecting the electron beam which is used for larger tubes, particularly for CRTs in computer monitors, radar, and television uses. A beam of electrons is a current flowing through a vacuum, and magnets will act on this current, deflecting the beam. The easiest way of doing this is to place coils around the neck of the tube and pass current through these coils to control the beam position on the face of the tube. This magnetic deflection method (Figure 8.10) is better suited for large CRTs such as were for many years used for television, monitors, or radar.

Figure 8.10:
Using magnetic deflection, with coils wrapped around the neck of the tube. This form of deflection allows very short tubes to be constructed

Summary

The CRT can use either electrostatic or magnetic deflection, so that the beam of electrons can have its direction altered, allowing the light spot to appear anywhere on the face of the tube. Magnetic deflection has been used for large televisions, computer monitors, and radar tubes; electrostatic deflection for the smaller tubes for measuring instruments. Nowadays, all the applications use flat-screen displays instead of a CRT.

The form of side-to-side deflection that is most common for CRTs is a **linear sweep**. This means that the beam is taken across the screen at a steady rate from one edge, and is returned very rapidly (an action called **flyback**) when it reaches the other edge. To generate such a linear sweep, a sawtooth waveform (Figure 8.11) is needed. An electrostatic tube can use a sawtooth voltage waveform applied to its deflection plates, and a magnetic deflection can use a sawtooth current applied to its deflection coils. The difference is important, because the electrostatic deflection requires only a sawtooth voltage with negligible current flowing, but the magnetically deflected tube requires a sawtooth **current**, and the voltage across the deflection coils will not be a sawtooth, because the coils act as a differentiating circuit. In fact, the voltage waveform is a pulse, and this is used in television receivers to generate a very high voltage for the CRT (see later, this chapter).

Liquid Crystal Display and Plasma Screens

Vacuum CRTs are now almost obsolete, for displays can now be made using LCDs and plasma principles. The principle of LCDs is that some liquid materials (including the cholesterol that clogs up your arteries) respond to the electric field between charged plates, and line up so that one end of each large molecule points towards the positive plate and the other end to the negative plate. In this condition, light passing through the material is polarized, so that if you look at the material through Polaroid™ sheet, the amount of light that passes alters when you

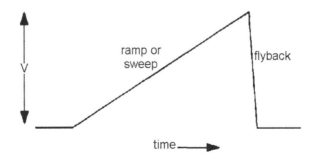

Figure 8.11:
The sawtooth or sweep wave that is needed to scan a cathode-ray tube. For an electrostatically deflected tube, this would be a voltage waveform applied to plates; for a magnetically deflected tube this is a current waveform applied to coils

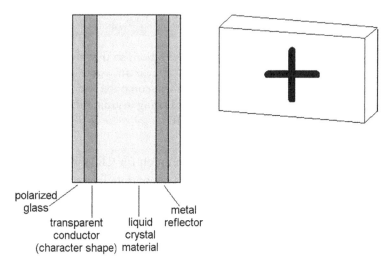

Figure 8.12:
Simplified cross-section of a liquid crystal display cell

rotate the Polaroid sheet. Polaroid is the material that is familiar from sunglasses that can cut out unwanted glare from water.

An LCD uses a set of parallel plates, of which one is a transparent conducting material, and one wall of the display is made of polarizing material (Figure 8.12). With no voltages applied, light passes through the polarized front panel, through the liquid, and is reflected from the metal backplate and out again. When a voltage is applied, the liquid between plates is polarized, and light can no longer be reflected back, making that part of the display look black. This type of display has been used for many years for calculators because it consumes very little power and yet gives a very clear output in good lighting conditions. It needs a source of light, however, either transmitted (light behind the screen) or reflected (light striking the front of the screen).

■ **Note**

The shapes that are to be produced are determined by the shape of the film of transparent conducting material at the front of the cell. For example, if this film is deposited in the shape of a '2', the cell will show this digit when activated.

■

For use as a display for portable computers, each position on a screen must be capable of being individually controlled, making the construction of such a screen much more difficult than that of a calculator display. For example, the usual older type of computer screen is required to be able to display at least 640×480 dots, a total of 307,200 dots; and that means 307,200 tiny LCD cells, each of which can be set to black or clear. For color displays, used almost universally now, dots are grouped in threes, one dealing with red, one with green, and one with blue. LCDs of this type have to be combined with integrated circuits to make the connections usable, and the later thin-film transistor (TFT) types are active, meaning that one or more integrated transistors will control each dot. Rather than depending on light reflection, these displays are made with a low-consumption backlight that shines through the clear portions of the display.

The technology of constructing LCD screens has now evolved to thin color screens with dot patterns of 1280 × 1024 or more, and these have now replaced the older CRT types of displays in computer monitors. Television receivers are also using these types of screen, and at the time of writing most television receivers offer HD, typically 1280 × 1024. A recent development is the use of light-emitting diodes (LEDs) to provide the backlight.

Plasma displays use a very different principle. Each dot in a plasma screen is produced by a tiny glass tube containing a gas and two electrical connections. A voltage between these connections will make the gas glow, and by specifying the appropriate gas in each tube we can produce the three basic colors of red, green, and blue which, when illuminated in the correct proportions, will produce white light (or any color). This system is best applied to large screens in which the larger dots are more appropriate; it is difficult to make in smaller sizes with high definition. One considerable advantage is that it needs no backlight, because the backlight for an LCD screen requires a considerable amount of power unless the LED system is used.

In the lifetime of this book you can expect to see several other forms of display technology such as organic light-emitting diode (OLED), laser phosphor display (LPD), trichromatic reflective electronic display (Tred), light-emitting polymer (LEP), and others that are still, at the time of writing, in the very early stages of development. Which of these we shall be using depends very much on factors such as ease of manufacture, lifetime, and price.

Analog Television Receiver

An analog television receiver for monochrome (Figure 8.13) has to deal with several signal-separating actions. The signal from the antenna is processed in a straightforward superhet type of receiver, and the methods that are used differ only because of the higher frequencies and the larger bandwidth. The vision and sound carriers are received together, and so the intermediate-frequency (IF) stages need to have a bandwidth that is typically about 6 MHz, enough to take the wideband video IF and the sound IF together. The real differences start at the demodulation block. A typical IF range is 35–39.5 MHz (for amplitude 6 dB down from maximum).

■ Note

For a digital television system, the superhet portion is the same as it would be for an analog receiver, but the processing of signals after the demodulation block is very different.

■

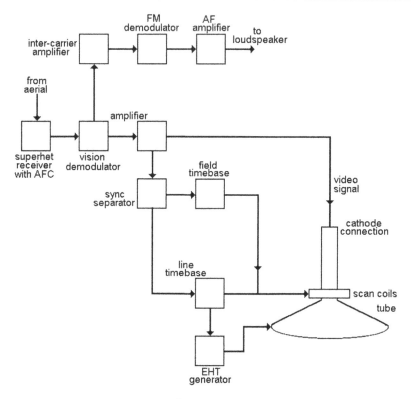

Figure 8.13:

A monochrome analog receiver block diagram. The diagram shows all of the superhet circuits as one block, and concentrated on the vision (luminance) and sound (audio) signals

Figure 8.14 illustrates the IF response, which must be broad enough to include both sound and vision carriers and sidebands, yet provide enough filtering to exclude signals from the adjacent frequencies. These adjacent frequencies arise from the mixing of the superhet oscillator frequency with the signals from other transmitters.

■ **Note**

Figure 8.14 also shows the frequency that is used as the sub-carrier for color signals (see later).

■

The demodulator is an amplitude demodulator, and at this stage the composite video signal with its sync pulses can be recovered. The effect of this stage on the IF for the sound signal is to act as a mixer, and since the sound IF and the vision IF are 6 MHz apart, another output of the demodulator is a frequency-modulated 6 MHz signal (the **intercarrier** signal) which carries the sound. This is separated by a filter, further amplified and (FM) demodulated to provide the sound output.

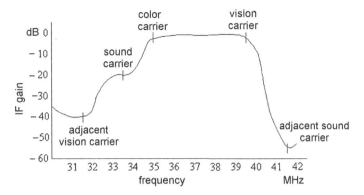

Figure 8.14:
The intermediate-frequency (IF) response of a typical receiver, showing the relative signal strengths over the bandwidth. The frequency marked as color carrier is important for color receivers

The video signal is amplified, and the synchronizing pulses are separated from it by a selective amplifier. These pulses are processed in a stage (the **sync separator**) which separates the line and the field pulses by using both differentiating and integrating circuits (Figure 8.15). The differentiated line pulses provide sharp spikes that are ideal for synchronizing the line oscillator, and the field pulses build up in the integrator to provide a field synchronizing pulse. The effect of the differentiator on the field pulses is ignored by the receiver because the beam is shut off during this time, and the effect of the line pulses on the integrator is negligible because these pulses are too small and too far apart to build into a field pulse.

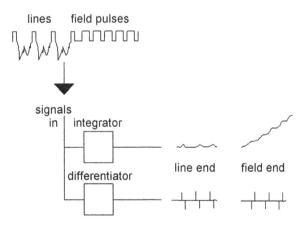

Figure 8.15:
How sync pulses are separated by integrating and differentiating circuits. The output of the differentiator during the field interval is ignored

■ Note

These pictures of sync pulses are simplified. At the end of a field or frame, there are five pulses called **equalizing pulses** placed before and following the field synchronizing pulses. The aim of this is to give time for the circuits to adjust to the differences between the end of a field and the end of a complete frame (there is half-a-line difference in timing).

■

The synchronized line and field timebase oscillators drive the output stages that deflect the electron beam, using magnetic deflection (passing scan currents through coils). The fast flyback of the line scan current causes a high voltage across the transformer that is used to couple the line scan output to the deflection coils, and this voltage is used to generate the extra-high-tension (EHT) supply of around 14–24 kV that is needed to accelerate the beam.

Meantime, the video signal has been amplified further and is taken to one terminal (cathode) of the CRT, with another terminal (grid) connected to a steady voltage to control brightness. Another set of circuits, the power supply unit (PSU), uses the alternating current (AC) mains supply to generate the steady voltages that will be used by the receiver circuits.

Summary

A monochrome receiver uses the normal superhet circuit up to the vision demodulator, where the composite video signal is recovered and the 6 MHz intercarrier sound signal is filtered off, demodulated, and amplified. The composite video signal is passed to a sync separator which removes the video portion and allows the two sets of synchronizing pulses to be separated and used to synchronize the timebases. The video signal is applied to the grid (or cathode) of the CRT, and the timebase signals to deflection coils. The very large pulses that exists on the line timebase output are stepped up by a transformer and used to generate the EHT supply of 14 kV or more for the CRT.

Color Television

Though Sam Goldwyn is reputed to have said that he would believe in color television when he saw it in black and white, the idea of color television is not new, and methods for transmitting in color have been around as for as long as we knew that black-and-white (monochrome) television was possible. Color television, like color printing and color photography (both demonstrated in the 1880s), relies on the fact that any color seen in nature can be obtained by mixing three primary colors. For light, these colors are red, green, and blue, and the primary colors used by painters are the paint colors (red, yellow, and blue) that absorb these light colors. To obtain color television, then, you must display together three pictures, one consisting of red light, one of blue light, and one of green light. This implies that the color television camera must generate three separate signals from the red, blue, and green colors in an image.

All of the early color television systems worked on what is called a **sequential** system, meaning that the colors were neither transmitted nor seen at the same time, relying on the optical effect called persistence of vision. This means that the eye cannot follow rapid changes, so that showing the red components of a picture followed rapidly by the blue and the green will appear to the eye as a complete color picture rather than flashing red, blue, and green. We rely on the same effect to fool the eye into believing that we are watching a moving picture rather than a set of still images.

A typical early method was the **frame sequential** system. Each picture frame was transmitted three times, using a different color filter for each of a set of three views so that though what was transmitted was black and white, each of a set of three frames was different because it had been shot through a color filter, one for each primary color. At the receiver, a large wheel was spinning between the viewer and the television screen, and this wheel carried a set of color filters. The synchronization was arranged so that the red filter would be over the CRT at the time when the frame containing the red image was being transmitted, so that this filter action put the color into the transmitted monochrome picture. The main snag with this system is that the frames must be transmitted at a higher rate to avoid flickering, and there are also problems with compatibility and with the synchronization of the wheel (and its size). A more realistic option was a line-sequential system, with each line being shown three times, once in red, once in green, and once in blue.

Several early television systems were devised to show still color pictures, but the first commercially transmitted color television signals were put out in 1948 by CBS in New York, using a combination of electronic and mechanical methods. The system was not successful, and a commission on color television decided that no scheme could be licensed unless it was **compatible**. In other words, anyone with a monochrome receiver had to be able to see an acceptable picture in black and white when watching a color broadcast, and anyone with a color receiver had to be able to see an acceptable black-and-white picture when such a picture was being transmitted (to allow for the use of the huge stock of black-and-white films that television studios had bought). This ruled out the use of frame-sequential systems, and almost every other consideration ruled out the use of mechanical systems allied to the CRT.

Summary

Color television, like color photography, depends on detecting the red, green, and blue light amplitudes in an image, since all natural colors can be obtained from mixtures of these pure light colors. A color television camera must therefore produce three separate signals, which can be R, G, and B, or mixtures from which separate R, G, and B signals can be obtained. Though early television systems had been able to produce pictures using sequential color, these systems were not compatible with monochrome transmissions and were abandoned in favor of a simultaneous system in 1952.

Radio Corporation of America (RCA) had been working on simultaneous systems throughout the 1940s, and their demonstrations in 1952 convinced the commission, the National Television Standards Committee (NTSC), that the RCA system was suitable. All other systems have taken this system as a basis, but differences in details mean that NTSC television pictures are not compatible with the PAL color system used in Germany, the UK, and other parts of western Europe, or the SECAM system used in France, former French colonies, and eastern Europe. This is also why you cannot exchange video cassettes or DVDs with friends overseas, and the hope that digital systems might be compatible has been dashed because the method of coding sound used in the USA is not the same as that used in Europe, though the vision signals are compatible. Even DVDs have to be separated into PAL, NTSC, and SECAM groups.

Let's start with the portions that are common to all analog systems. To start with, the image is split into red, green, and blue signal components, using prisms, so that three separate camera tubes or semiconductor CCD detectors can each produce a signal for one primary color. This is the starting point for the block diagram of Figure 8.16. The signals are mixed to form three outputs. One is a normal monochrome signal, called the **Y signal**. This is a signal that any monochrome receiver can use and it must be the main video signal so as to ensure compatibility. The other two outputs of this mixer are called **color-difference** signals, obtained by using different mixtures of the red, green, and blue signals. These color-difference signals are designed to make the best use of the transmission system and are of considerably lower

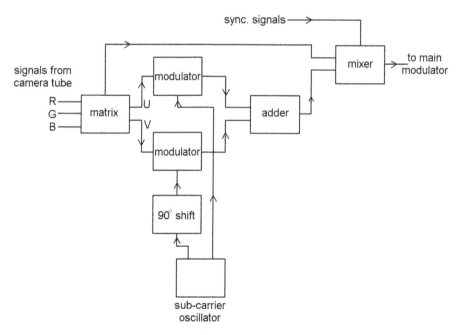

Figure 8.16:
The PAL system. A block diagram, simplified, of the processes at the transmitter

amplitude than the monochrome signal, because in any television signal, the monochrome portion always carries much more information. Color is less important, and all the fine details of a picture can be in monochrome only because the eye is less sensitive to color in small areas. These color-difference signals are lettered **U** and **V**.

The color-difference signals are transmitted using a **sub-carrier**, a method that we have looked at already in connection with stereo radio. This time, the use of the sub-carrier is much more difficult because it has to carry more than one signal. To ensure compatibility, the sub-carrier frequency must be within the normal bandwidth of a monochrome signal, so that you would expect it to cause a pattern on the screen. As it happens, by choosing the sub-carrier frequency carefully, and keeping its amplitude low, it is possible to make the pattern almost invisible. The other problem is that this sub-carrier has to be modulated with **two** signals, and this is done by modulating one color-difference signal on to the sub-carrier directly (using amplitude modulation) and then using a phase-shifted version of the sub-carrier, with phase shifted by 90°, and modulating the other color-difference signal on to this phase-shifted sub-carrier.

■ **Note**

Modulating two waves of the same frequency but with 90° phase difference is equivalent in effect to modulating both the amplitude and the phase of the same carrier. It is just like the way that stereo recordings vibrate a stylus in directions at right angles to each other.

■

If you remember that a 90° phase shift means that one wave is at zero when the other is at a maximum, you will appreciate that adding these two modulated sub-carriers together does not cause them to interfere with each other. The doubly modulated sub-carrier can now be added to the monochrome signals (remember that the sub-carrier has a considerably lower amplitude) and the synchronizing signals are added. This forms the color composite video waveform which can be modulated on to the main carrier. What this amounts to is that the amplitude of the modulated signal represents the saturation of color and the phase of the modulated signal represents the hue.

Definition

The **hue** is the color, the wavelength of the light, and **saturation** measures how intense the hue is. Most natural colors have low saturation values, meaning that these colors are heavily diluted with white.

Synchronizing is not just a matter of the line and field pulses this time. A color receiver has to be able to generate locally a copy of the sub-carrier in the correct phase so that it can demodulate the color signals. To ensure this, ten cycles of the sub-carrier are transmitted along

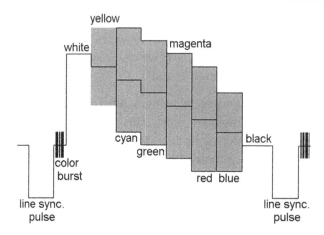

Figure 8.17:
The color burst and color bars. The color burst consists of ten cycles (actually 10 ± 1 cycles) of sub-carrier located in the back porch of each line-sync pulse. The diagram also shows the waveforms corresponding to color bars; this cannot, of course, indicate the phase of the sub-carrier

with each line-synchronizing pulse, using a time when the normal monochrome signal has a gap between the line-synchronizing pulse and the start of the line signal (Figure 8.17). This **color sync** interval is called the **back porch** of the synchronizing pulse. The illustration also shows how some colored bars appear on a signal, as viewed by an oscilloscope (see Chapter 17). The sub-carrier in these signals is represented by shading. If the sub-carrier were not present, these bars would appear in shades of gray.

Summary

To ensure compatibility, the normal monochrome signal must be modulated on to the main carrier in the usual way. The color-difference signals are then transmitted by modulating them on to a sub-carrier whose frequency is carefully chosen to cause the minimum of interference with the monochrome signal. One color-difference signal is amplitude-modulated on to the sub-carrier directly, and the other is amplitude modulated on to a sub-carrier of the same frequency but phase shifted by 90°. These signals are added to the monochrome signal, and nine cycles of the 0° phase sub-carrier signal are inserted following each line-synchronizing pulse so that the receiver can locally generate a sub-carrier in the correct phase.

The Three Color Systems

Compatible analog color television broadcasting started in the USA in 1952, using a scheme that was very much as has been outlined here (though the color difference signals were not formed in the same way). In the early days the NTSC system suffered from color problems, with viewers complaining that they constantly needed to adjust the color controls that were fitted to receivers, hence the old joke that NTSC stood for 'never twice the same color'. The problem was that during transmission of the signals, changes in the phase of the carrier caused by reflections had a serious effect on the sub-carrier, causing the color information to alter. This is because the phase of the color signal carried the hue information. Though subsequent development has greatly reduced these problems, the basic NTSC system has remained virtually unaltered.

By contrast, the color television systems used in Europe were designed much later with an eye on the problems that had been experienced in the USA. Though the principle of transmitting a monochrome signal along with a sub-carrier for two color signals has not changed, the way that the color sub-carrier is used has been modified, and so too has the composition of the color signals. There are two main European analog systems, PAL and SECAM. The PAL system was evolved by Dr Bruch at Telefunken in Germany, and the SECAM system by Henri de France and a consortium of French firms. SECAM is used in France, in former French colonies, and in eastern Europe, but the PAL system has been more widely adopted. Only the American-influenced countries (North and South) and Japan retain the modern form of the NTSC system.

The PAL system uses the color-difference signals that are called U and V, with the color mix carefully chosen so that these signals need only a small (and equal) bandwidth. What makes the essential difference, however, is that the V signal, which carries most of the hue information, is inverted on each even-numbered line, and the signal that is used in the receiver is always the average of one line and the preceding line. If there has been a phase shift in the sub-carrier caused by conditions between the transmitter and the receiver, then subtracting the V signals of one line from the V signals of the following line will have the effect of canceling out the changes. Since the V signals carry the hue information, this eliminates the changes of color that were such a problem with the original NTSC signals. The snag is that the averaging of adjacent lines has the effect of reducing the vertical resolution for color to half of its value for monochrome, but, since the eye is less sensitive to color, this is not as much of a problem as it might appear.

By contrast, the SECAM system works by using frequency modulation of the sub-carrier, using the U signal on one line and the V signal on the next. As for the PAL system, the information of two lines has to be gathered up to provide the signals for each one line, reducing the vertical resolution for color.

Summary

The European analog color television systems have been designed from the start to avoid the problems caused by phase changes of the color signal. The PAL system does this by inverting the V signal on alternate lines and averaging the signals at the receiver to cancel out the effects of phase change. SECAM operates by transmitting U and V signals on alternate lines. In both cases, the color information in two successive lines is always averaged.

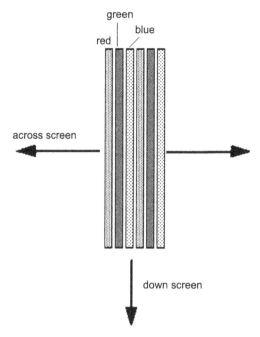

Figure 8.18:
The arrangement of phosphor stripes on the screen of a color display tube. Each dot viewed on the tube face consists of portions of a set of three stripes

Color Television Tubes

The first color television receivers depended very heavily on the CRT, and the type that was universally used in the latter days was the color-stripe type. This replaced the color-dot type that was used in 1952 and continued in use until the late 1960s. These tubes allow simultaneous color output, meaning that the colors of a picture are all being displayed together rather than in a sequence, so that color television tubes have to use three separate electron guns, one for each primary color.

Figure 8.18 shows a magnified view of a portion of the screen of a typical tube. The glowing phosphors are arranged as thin stripes, using three different materials that glow each in a primary color when struck by electrons. These stripes are narrow, and typically a receiver tube would use at least 900 stripes across the width of the screen. Tubes for computer monitors used a much larger number, and this was reflected in the price of a monitor compared to a television receiver of the same screen size. CRT monitors are by now seldom used because the size of screen and the requirement for high resolution made the use of LCD flat panels economically possible.

The color tube contains three separate electron guns (Figure 8.19). Between the screen and the electron guns and close to the screen there is a metal mask (formerly called the **shadowmask**

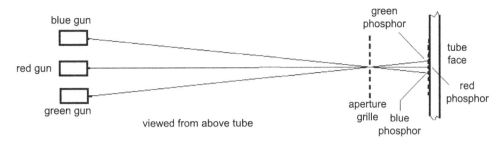

Figure 8.19:
How the three guns of a color tube are arranged. The aperture grille ensures that electrons from the 'red' gun strike only red-glowing phosphor stripes, and similarly for the other two guns

when round apertures were used, later called the **aperture grille**) consisting of narrow slits. These slits are lined up so that the electrons from one gun will strike only the phosphor stripes that glow red, the electrons from the second gun will strike only the stripes that glow green, and the electrons from the third gun will hit only the blue stripes. We can therefore call the electron guns **red**, **green**, and **blue** because these are the colors that they will produce on the screen. If we can obtain a set of video signals for these guns, identical to the original red, green, and blue signals from the camera tubes, we should achieve an acceptable copy of the television image in color.

■ **Note**

Another approach to display is the use of projection displays, using either small (about 7 inches) CRTs, or LCD panels with bright illumination. These displays can be very impressive, though some types need to be viewed in semi-darkness for best results.

■

Summary

The display tube is the heart of any CRT color television receiver. All modern color CRTs use a set of phosphor strips, arranged in a recurring R, G, B pattern. Close to this screen, a metal mask ensures that each of the three electron guns will project a beam that can hit only one color strip each, so that signals to these guns will be R, G, and B signals, respectively. The use of LCDs has advanced so rapidly that few (if any) television receivers using a CRT are now available.

Analog Receiver Circuits

A fair amount of the circuitry of an analog color television receiver is identical to that of a monochrome receiver. The superhet principle is used, and the differences start to appear only following the vision detector, where the video signal consists of the monochrome signal together with the modulated sub-carrier. The U and V color-difference signals have to be recovered from this sub-carrier and combined with the monochrome signal to give three separate R, G, and B signals that can be used on the separate guns of the color tube.

A simplified outline of a receiver, showing the usual superhet stages as one block and neglecting differences between systems, is shown in Figure 8.20. The video signal has the 6 MHz sound signal filtered off, and the vision signal is amplified. At the output of this amplifier, other filters are used to separate the sub-carrier signal, which (in the UK) is at 4.43 MHz, with its sidebands, and the monochrome or **luminance** signal, with the sub-carrier

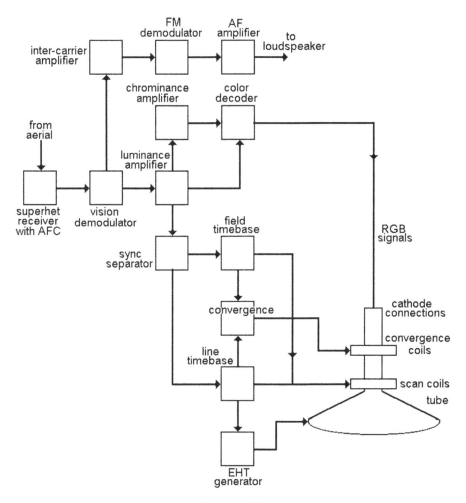

Figure 8.20:
A general block diagram for a color receiver. The convergence circuits keep the individual electron beams aimed at their respective phosphor stripes. Without convergence corrections, the picture is satisfactory only at the middle of the screen

frequencies greatly reduced by filtering, is separated off and amplified. The sub-carrier (**chrominance**) signal has to be demodulated in two separate circuits, because we have to recover two separate signals, U and V, from it. This requires a circuit in which the unmodulated sub-carrier frequency, in the correct phase, is mixed with the modulated sub-carrier signal. This type of circuit is called a **synchronous demodulator**.

A pulse is taken from the line timebase and used to switch on gates, circuits that will pass signal only for a specified interval. One of these gates allows the color-synchronizing 'burst' to pass to the local sub-carrier oscillator to maintain it at the correct frequency and phase. This oscillator output is fed to a demodulator whose other input is the modulated sub-carrier, and this has the effect of recovering the V signal. The oscillator output is also phase shifted through 90° and this shifted wave is used in another demodulator to recover the U signal. The U, V, and Y (monochrome) signals are then mixed to get the R, G, and B signals that are fed to the CRT guns.

That is the simplest possible outline, and it is quite close to the original NTSC system for receivers. The PAL receiver incorporates more complications because at the camera end the V signal was inverted on each line, and the receiver needs to be able to combine the V signals from each pair of lines by storing the information of each line and combining it with the following line.

This is done using the block diagram of Figure 8.21, which shows part of the PAL receiver concerned with color decoding. One output from the crystal oscillator is passed to an inverter circuit, and this in turn is controlled by a bistable switch. The word bistable, in this context, means that the switch will flip over each time it receives a pulse input, and its pulse inputs are from the line timebase of the receiver, so that the switch is operated on each line. The action of the switch on the inverter will ensure that the oscillator signal is in phase on one line and will reverse (180° phase shift) on the next line, and so on. Also shown in this diagram is an arrangement for an identification (**ident**) signal, taken from the burst signal, that makes certain that the bistable switch is itself operating in phase, and not inverting a signal that ought to be unchanged.

The sub-carrier signals are fed to a time delay. This is arranged to delay the signal by exactly the time of one line, and the conventional method originally was to use a glass block. The electrical signals were converted into ultrasonic soundwave signals, and they traveled through the glass to a pickup where they were converted back into electrical signals. The delay is the time that these ultrasound signals take to travel through the glass block, and the dimensions of the glass are adjusted so that this is exactly the time of a line. More modern receivers used electronic digital delay circuits, using the principle of the shift register (see Chapter 11).

The delayed signal (the sub-carrier for the previous line) is added to the input signal (the sub-carrier for the current line) in one circuit and subtracted in a second circuit so as to produce

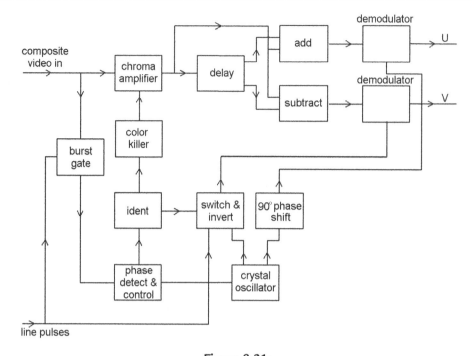

Figure 8.21:
The decoding of the PAL type of signal. This is a very complicated process compared with the older NTSC system

averaged signals for the U and V demodulators that will contain no phase errors. Note that averaging by itself would not correct phase changes; it is the combination of averaging with the phase reversal of the V signals that ensures the correction.

In addition, there are embellishments on the circuits, and one of these is the color **killer**. When the transmitted signals are in monochrome, any signal at the sub-carrier frequency would produce color effects at the receiver, so that these circuits must be switched off for transmissions (old films, for example) that contain no color. This is done by using the color burst signals, so that when a color burst is present, a steady voltage from a demodulator will ensure that the color circuits are biased on and working. In the absence of the burst signal, the color circuits are biased off, ensuring that any frequencies around 4.33 MHz (the color sub-carrier frequency for the UK) in the picture signal do not cause colors to appear.

Summary

The color processing for an analog receiver consists of separating off the signals at the sub-carrier frequency. A copy of the sub-carrier is generated and its phase and frequency are corrected by the burst signal. This sub-carrier can be used in demodulators to recover the original U and V color difference signals, and these can be combined with the luminance signal to form separate R, G, and B color signals. For the PAL system, the sub-carrier used for the V signal must be inverted on every other line, and an ident signal is used to ensure that the inversions are in step. In addition, the color signals are passed through a time delay of exactly the time of one line, so that the color signals of one line can be combined with those of the previous line (with the V signal inverted) to provide averaged U and V signals.

We will look at the digital television systems that have completely replaced the analog system in Chapter 16.

CHAPTER 9

Digital Signals

Voltage Levels

Definition

A digital signal is one in which a change in voltage, and the time at which it occurs, are of very much more importance than the precise *size* of the change or the exact *shape* of the waveform.

All of the waveforms used in digital circuits are steep-sided pulses or square waves and it is the *change* in voltage that is significant, not the *values* of voltage. For that reason, the voltages of digital signals are not referred to directly, only as 1 and 0. The important feature of a digital signal is that each change is between just two voltage levels, typically 0 V and +5 V, and that these levels need not be precise. In this example, the 1 level can be anything from 2.4 V to 5.2 V, and the 0 level anything from 0 V to +0.8 V. We could equally well define 1 as +15 V and 0 as −15 V. By using 0 and 1 in place of the actual voltages, we make it clear that digital electronics is about levels (representing numbers), not waveforms.

The importance of using just two digits, 0 and 1, is that this is ideally suited to using all electronic devices. A transistor, whether a bipolar or field-effect transistor (FET) type, can be switched either fully on or fully off, and these two states can be ensured easily, much more easily than any other states that depend on a precise bias. By using just these two states, then, we can avoid the kind of errors that would arise if we tried to make a transistor operate with, say, ten levels of voltage between two voltage extremes. By using only two levels, the possibility of mistakes is made very much less. The only snag is that any counting that we do has to be in terms of only two digits, 0 and 1, and counting is the action that is most needed in digital circuitry.

Counting with only two digits requires using a scale of two called the **binary scale**, in place of our usual scale of ten (a denary scale). There is nothing particularly difficult about this, because numbers in the conventional binary scale are written in the same way as ordinary (denary) numbers. As with denary numbers, the position of a digit in a number is important. For example, the denary number 362 means three hundreds, six tens, and two units. The position of a digit in this scale represents a power of 10, with the right-hand position (or **least significant**

Electronics Simplified. DOI: 10.1016/B978-0-08-097063-9.10009-3

position) for units, the next for tens, the next for hundreds (ten squared), then thousands (ten cubed), and so on.

For a scale of two, the same scheme is followed. In this case, however, the positions are for powers of two, as units, twos, fours (two squared), eights (two cubed), and so on. Table 9.1 shows powers of two as place numbers and their denary equivalents, and the text shows how a binary number can be converted to denary form and a denary number to binary form.

■ **Note**

In a binary number such as 1100, the last zero is the **least significant bit**, and the first 1 is the **most significant bit.** Arithmetic actions start with the least significant bit and work towards the most significant bit, shifting left through the digits.

■

Digital circuits are switching circuits, and the important feature is fast switching between the two possible voltage levels. Most digital circuits would require a huge number of transistors to construct in discrete form, so that digital circuits make use of integrated circuits (ICs), mostly of the MOSFET type, exclusively. The use of integrated construction brings two

Table 9.1: Two place numbers and denary equivalents.

Place no.	Denary	Place no.	Denary	Place no.	Denary	Place no.	Denary
0	1	8	256	16	65536	24	16777216
1	2	9	512	17	131072	25	33554432
2	4	10	1024	18	262144	26	67108864
3	8	11	2048	19	524288	27	134217728
4	16	12	4096	20	1048576	28	268435456
5	32	13	8192	21	2097152	29	536870912
6	64	14	16384	22	4194304	30	1073741824
7	128	15	32768	23	8388608	31	2147483648

Converting binary to denary: For each 1 in the number, write down the denary number for that place, then add.
Example: Binary 10001110110. This uses place number 0 to 10, with 1s in positions 1, 2, 4, 5, 6 and 10. This corresponds to denary numbers $1024 + 64 + 32 + 16 + 4 + 2 = 142$.
Converting denary to binary. Divide the number by 2, and write the remainder (which must be 0 or 1) at the side. Now divide the last result also by 2, again writing the remainder at the side. Repeat until the last remainder (which is always 1) has been found, then read the numbers from **bottom to top.**
Example: Denary 875 $875 \div 2 = 437$ remainder 1
$437 \div 2 = 218$ remainder 1
$218 \div 2 = 109$ remainder 0
$109 \div 2 = 54$ remainder 1
$54 \div 2 = 27$ remainder 0
$27 \div 2 = 13$ remainder 1
$13 \div 2 = 6$ remainder 1
$6 \div 2 = 3$ remainder 0
$3 \div 2 = 1$ remainder 1
$1 \div 2 = 0$ remainder 1
The binary number is therefore 1101101011.

particular advantages to digital circuits. One is that circuits can be very much more reliable than when separate components are used (in what are called discrete circuits). The other advantage is that very much more complex circuits, with a large number of components, can be made as easily in integrated form as simple circuits once the master templates have been made. The double advantage of reliability and cost is what has driven the digital revolution.

Because digital systems are based on counting with a scale of two, their first obvious applications were to calculators and computers, topics we shall deal with in Chapter 13. What is much less obvious is that digital signals can be used to replace the analog type of signals that we have become accustomed to. This is the point we shall pay particular attention to in this chapter, because some of the most startling achievements of digital circuits are where digital methods have completely replaced analog methods, such as the audio compact disc (CD) and the digital versatile disc (DVD), and in the digital television and radio systems that have replaced or are about to replace the analog systems that we have grown up with.

As it happens, the development of digital ICs has had a longer history than that of analog IC devices. When ICs could first be produced, the manufacturing of analog devices was extremely difficult because of the difficulty of ensuring correct bias and the problems of power dissipation. Digital IC circuits, using transistors that were either fully off or fully on, presented no bias problems and had much lower dissipation levels. In addition, circuits were soon developed that reduced the dissipation still further by eliminating the need for resistors on the chip. Digital ICs therefore had a head start as far as design and production were concerned, and because they were immediately put into use, the development of new versions of digital ICs was well under way before analog ICs made any sort of impact on the market.

Summary

Digital signals consist of rapid transitions between two voltage levels that are labeled 0 and 1 — the actual values are not important. This form of signal is well suited to active components, because the 0 and 1 voltages can correspond to full-on and cut-off conditions, respectively, each of which causes very little dissipation. ICs are ideally suited to digital signal use because complex circuits can be manufactured in one set of processes, dissipation is low, reliability is very high, and costs can be low.

Recording Digitally

Given, then, the advantages of digital signals as far as the use of transistors and ICs is concerned, what are the advantages for the processing of signals? The most obvious advantage relates to tape or any other magnetic recording. Instead of expecting the magnetization of the tape to reproduce the varying voltage of an analog signal, the tape magnetization will be either maximum in one direction or maximum in the other. This is a technique called **saturation recording** for which the characteristics of most magnetic recording materials are ideally suited. The precise amount of magnetization is no longer important, only its direction. This, incidentally, makes it possible to design recording and replay heads rather differently so that a greater number of signals can be packed into a given length of track on the tape. Since the precise amount of magnetization is not important, linearity problems disappear.

■ Note

This does not mean that **all** problems disappear. Recording systems cannot cope well with a stream of identical digits, such as 11111111111 or 0000000000, and coding circuits are needed to make sure that all the numbers that are recorded contain both 1s and 0s, with no long sequences of just one digit. This complication can be taken care of using a specialized IC.

■

Noise problems are also greatly reduced. Tape noise consists of signals that are, compared to the digital signals, far too small to register, so that they have no effect at all on the digitally recorded signals. This also makes digital tapes easier to copy, because there is no degradation of the signals caused by copying noise, as there always is when conventional analog recorded tapes are copied. Since linearity and noise are the two main problems of any tape (or other magnetic) recording system it is hardly surprising that recording studios have rushed to change over to digital tape mastering.

The surprising thing is that it has been so late in arriving on the domestic scene, because the technology has been around for long enough, certainly as long as that of videotape recording. A few (mainly Betamax) video recorders provided for making good-quality audio recordings of up to eight hours on videotape, but this excellent facility was not taken up by many manufacturers, and died out when VHS started to dominate the video market in the UK.

The advantages that apply to digital recording with tape apply even more forcefully to discs. The accepted standard (CD) method of placing a digital signal on to a flat plastic disc is to record each binary digit (bit) 1 as a tiny pit or bump on the otherwise flat surface of the disc, and interpret, on replay, a change of reflection of a laser beam as the digital 1. Once again, the exact size or shape of the pit/bump is unimportant as long as it can be read by the beam, and only the number of pits/bumps is used to carry signals. We shall see later that the process is by no means so simple as this would indicate, and the CD is a more complicated and elaborate system than the tape system (DAT) that briefly became available, though at prohibitive prices, in the UK. DAT is now just a dim memory for domestic recorders.

The basic CD principles, however, are simple enough, and they make the system immune from the problems of the long-playing (LP) disc. There is no mechanical cutter, because the pits have been produced by a laser beam that has no mass to shift and is simply switched on and off by the digital signals. At the replay end of the process, another (lower power) laser beam will read the pattern of pits or bumps and once again this is a process that does not require any mechanical movement of a stylus or any pickup mechanism, and no contact with the disc itself. Chapter 12 is devoted to the CD system and later developments in digital processing.

■ Note

DVD uses the same principles as CD, but crams more information on to the disc by using a shorter wavelength of laser light, two layers of recorded information, and in some cases both sides of the disc. We will look at the Blu-ray development of DVD later.

■

As with magnetic systems, there is no problem of linearity, because it is only the number of pits or bumps rather than their shape and size that counts. Noise exists only in the form of a miscount of the pits/bumps or as confusion over the least significant bit of a number, and as we shall see there are methods that can reduce this to a negligible amount. Copying of a CD is not quite so easy as the molding process for LPs, but it costs much less than copying a tape, because a tape copy requires winding all of the tape past a recording head, which takes much longer than a stamping action. CDs are therefore more profitable to manufacture than tapes.

A CD copy is much less easily damaged than its LP counterpart, and even discs that look as if they had been used to patch a motorway will usually play with no noticeable effects on the quality of the sound, though such a disc will sometimes skip a track. CD recording machines, mainly intended for computer CD-ROM, are now no more expensive than a tape-recording system. DVD recorders have now been available for some time, and can be used to replace the current generation of DVD players that have no recording facility.

Summary

Outside computing, digital systems are best known for the CD, DVD, Blu-ray, and to a lesser extent in magnetic recording. Virtually every recording now made uses digital tape systems for mastering, however, so that digital methods are likely to be used at each stage in the production of a CD (this is marked by a **DDD** symbol on a CD). Magnetic recording uses a saturation system, using one peak of magnetization to represent 1 and the opposite to represent 0. Disc recording uses tiny pits/bumps on a flat surface to represent a 1 and the flat surface to represent 0.

Digital Broadcasting

Digital broadcasting means transmitting a signal, usually sound or video, in digital form over an existing system such as cable, satellite, or the **terrestrial** (earth-bound) transmitters. We will look in more detail at digital broadcasting in Chapter 16, but for the moment, it is useful to consider some of the advantages.

First of all, digital broadcasting does not tie you to **real time**. When you broadcast in an analog system, the sound that you hear is what is being transmitted at that instant, allowing for the small delay caused by the speed of electromagnetic waves. Using digital transmission, you can transmit a burst of digital signals at a very fast rate, pause, transmit another burst, pause, and so on. At the receiver the incoming signals are stored and then released at the correct steady rate. This has several advantages:

- You can **multiplex** transmissions, sending a burst of one program, a burst of the next, and so on, and separate these at the receiver. We will see later that six television transmissions can be sent on one carrier frequency (using the normal 8 MHz bandwidth) in this way.
- You can send signals alternately instead of together. For example, the left and right stereo sound signals can be sent alternately as a stream of L, R, L, R ... codes.
- You need not send the digital codes in the correct order, provided that you can reassemble them at the receiver. This reduces interference problems because a burst of interference will not affect all the codes of a sequence if they are sent at different times.

■ **Note**

One odd effect is that if you watch an analog television and a digital television together, the sound and picture on the digital receiver lag behind the sound and picture on the analog receiver. This is because of the storage time in the digital receiver when the signals are being combined.

■

The most compelling advantage is that you can manipulate digital signals in ways that would be quite impossible with analog signals. This is why digital methods were being used in studio work well before any attempt was made to use them for broadcasting. Providing that a manipulation of digital signals can be reversed at the receiver, it has no effect on the final signal. By contrast, passing an analog signal through one filter and then through one with the opposite effect would affect the end result seriously.

Finally, converting a signal to digital form can allow you to reduce redundancy. An analog television picture of a still scene uses the transmitter to send the same set of signals 25 times each second. A digital version would send one set of signals and hold it in memory until the picture changed. In addition, digital manipulation of analog signals often results in other redundant pieces of code that can be eliminated.

There is one important disadvantage that made digital broadcasting seem an impossible dream until recently. Digitizing any waveform results in a set of digits that are pulses, repeating at a high speed and requiring a wide bandwidth for transmission. The CD system deals with this wide bandwidth, but it is not so simple for broadcasting because frequency allocations cannot easily be changed. The way round this is **digital compression**, removing redundancy in the data until the rate of sending bits can be reduced so far that a transmission will fit easily into the available bandwidth. As we have seen above, this can be done to such an extent that several digital television transmissions can be fitted into the bandwidth of a single analog signal.

■ Note

We shall look at compression methods later, but it is important to note here that the methods used for digital television and radio depend as much on knowledge of how the eye, ear, and brain interact as on the electronics systems. In other words, these methods depend to a considerable extent on knowing just how much information can be omitted without the viewer/listener noticing.

■

By using compression systems, digital sound and vision signals can be sent over normal radio/television systems. Digital signals can be manipulated to a much greater extent than analog signals without loss of signal content.

Conversions

No advantages are ever obtained without paying some sort of price, and the price to be paid for the advantages of digital recording, processing, and reproduction of signals consists of the problems of converting between analog and digital signal systems, and the increased rate of processing of data. For example, a sound wave is not a digital signal, so that its electrical counterpart must be converted into digital form. This must be done at some stage where the electrical signal is of reasonable amplitude, several volts, so that any noise that is caused will be negligible in comparison to the signal amplitude. That in itself is no great problem, but the nature of the conversion is.

What we have to do is to represent each part of the wave by a number whose value is proportional to the voltage of the waveform at that point. This involves us right away in the two main problems of digital systems: resolution and response time. Since the conversion to and from sound waves is the most difficult challenge for digital systems, we shall concentrate on it here. By comparison, radar, and even television, are systems that were almost digital in form even from the start. For example, the television line waveform consists of the electrical voltage generated from a set of samples of brightness of a line of a scanned image, and it is as easy to make a digital number to represent each voltage level as it is to work with the levels as a waveform.

To see just how much of a problem the conversion of sound waves is, imagine a system that used only the numbers -2 to $+2$, on a signal of 4 V total peak-to-peak amplitude. If this were used to code a waveform (shown as a triangular wave for simplicity) as in Figure 9.1(a) then since no half-digits can exist, any level between -0.5 V and $+0.5$ V would be coded as 0, any signal between $+0.5$ V and $+1.5$ V as 1 and so on, using ordinary denary numbers rather than

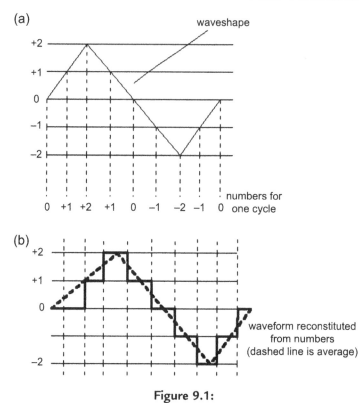

Figure 9.1:
Quantizing a waveform. Each new level of voltage is represented by the number for that level, so that the waveform is coded as a stream of numbers. Reversing the process produces a shape that when smoothed (averaged) provides a recognizable copy of the input even for this very crude five-level system

binary numbers here to make the principle clearer. In other words, each part of the wave is represented by an integer (whole) number between −2 and +2, and if we plotted these numbers on the same graph scale then the result would look as in Figure 9.1(b). This is a 'block' shape of wave, but recognizably a wave which if heavily smoothed would be something like the original one.

We could say that this is a five-level quantization of the wave, meaning that the infinite number of voltage levels of the original wave has been reduced to just five separate levels. This is a very crude quantization, and the shape of a wave that has been quantized to a larger number of levels is a much better approximation to the original. The larger the number of levels, the closer the wave comes to its original pattern, though we are cheating in a sense by using a sinewave as an illustration, since this is the simplest type of wave to convert in each direction; we need know only one number, the peak amplitude, to specify a sinewave. Nevertheless, it is clear that the greater the number of levels that can be expressed as different numbers then the better is the fidelity of the sample.

Definition

Quantization means the sampling of a waveform so that the amplitude of each sample can be represented by a number. It is the essential first step in converting from analog form to digital form.

In case you feel that all this is a gross distortion of a wave, consider what happens when an audio wave of 10 kHz is transmitted by medium-wave radio, using a carrier wave of 500 kHz. One audio wave will occupy the time of 50 radio waves, which means in effect that the shape of the audio wave is represented by the amplitudes of the peaks of 50 radio waves, a 50-level quantization. You might also like to consider what sort of quantization is involved when an analog tape system uses a bias frequency of only 110 kHz, as many do. Compare this with the 65,536 levels used for a CD.

The idea of carrying an audio wave by making use of samples is not in any way new, and is inherent in amplitude modulation (AM) radio systems which were considered reasonably good for many years. It is equally inherent in frequency modulation (FM), and it is only the use of a fairly large amount of frequency change (the peak deviation) that avoids this type of quantization becoming too crude. Of all the quantized ways of carrying an audio signal, in fact, FM is probably the most satisfactory, and FM methods are often adopted for digital recording, using one frequency to represent a 0 and another to represent a 1. Another option that we shall look at later is changing the phase and amplitude of a wave, with each different phase and amplitude representing a different set of digital bits.

Summary

The conversion of a waveform into a set of digital signals starts with quantization of the wave to produce a set of numbers. The greater the number of quantization levels, the more precise the digital representation, but excessive quantization is wasteful in terms of the time required.

This brings us to the second problem, however. Because the conversion of an audio wave into a set of digits involves sampling the voltage of the wave at a large number of intervals, the digital signal consists of a large set of numbers. Suppose that the highest frequency of audio signal is sampled four times per cycle. This would mean that the highest audio frequency of 20 kHz would require a sampling rate of 80 kHz. This is not exactly an easy frequency to record even if it were in the form of a sinewave, and the whole point of digital waveforms is that they are not sinewaves but steep-sided pulses which are considerably more difficult to record. From this alone, it is not difficult to see that digital recording of sound must involve rather more than analog recording.

The next point is the form of the numbers. We have seen already that numbers are used in binary form in order to allow for the use of only the two values of 0 and 1. The binary code that has been illustrated in this chapter is called 8-4-2-1 binary, because the position of a digit represents the powers of two that follow this type of sequence. There are, however, other ways of representing numbers in terms of 0 and 1, and the main advantage of the 8-4-2-1 system is that both coding and decoding are relatively simple. Whatever method is used, however, we cannot get away from the size that the binary number must have. It is generally agreed that modern digital audio for music should use a 16-bit number to represent each wave amplitude, so that the wave amplitude can be any of up to 65,536 values. For each sample that we take of a wave, then, we have to record 16 digital signals, each 0 or 1, and all 16 **bits** will be needed in order to reconstitute the original wave. We refer to this as a **bit-depth** of 16. Bit-depths higher than this are used for professional equipment; a bit-depth of 24 is typical.

Definition

A bit is short for a binary digit, a 0 or 1 signal. By convention, bits are usually gathered into a set of eight, called a **byte**. A pair of bytes, 16 bits, is called a **word**. Unfortunately, the same term is also used for higher number groupings such as 32, 64, 128, etc.

Digital Coding

The number of digital signals per sample is the point on which so many attempts to achieve digital coding of audio have foundered in the past. As so often happens, the problems could be more easily solved using tape methods, because it would be quite feasible to make a 16-track tape recorder using wide tape and to use each channel for one particular bit in a number. This is, in fact, the method that can be used for digital mastering where tape size is not a problem, but the disadvantage here is that for original recordings some 16—32 separate music tracks will be needed. If each of these were to consist of 16 digital tracks the recorder would, to put it mildly, be rather overloaded. Since there is no possibility of creating or using a 16-track disc, the attractively simple idea of using one track per digital bit has to be put aside. The alternative is **serial** transmission and recording.

Definition

Serial means one after another. For 16 bits of a binary number, serial transmission means that the bits are transmitted in a stream of 16 separate signals of 0 or 1, rather in the form of separate signals on 16 channels at once. The rate of transmission of a digital serial signal is stated in kb/s (kilobits per second) or Mb/s (megabits per second).

Now if the signals are samples that have been taken at the rate of 40 kHz, and each signal requires 16 bits to be sent out, then the rate of sending digital signals is 16×40 kb/s, which if we coded it as one wave per bit would be equivalent to a frequency of 640 kHz, well beyond the rates for which ordinary tape or disc systems can cope. As it happens, we can get away with much more efficient coding and with slower sampling rates, as we shall see, but this does not offer much relief because there are further problems.

■ Note

A well-known rule (Shannon's law, which was worked out long before digital recording was established) states that the sampling rate must be at least twice the highest component of the analog signal, so that a 40 kHz sampling rate is adequate for a signal whose highest frequency is 20 kHz. Remember that the components of an audio signal that are at this highest frequency will be present only in very small amplitudes.

■

When a parallel system is used, with one channel for each bit, there is no problem of identifying a number, because the bits are present at the same time on the 16 different channels, with each bit in its place; the most significant bit will be on the MSB line, the least significant on the LSB line, and so on. When bits are sent one at a time, though, how do you know which bits belong to which number? Can you be sure that a bit is the last bit of one number or is it the first bit of the next number? The point is very important because when the 8-4-2-1 system is used, a 1 as the most important bit means a value of 32,768, but a 1 as the least important bit means just 1. The difference in terms of signal amplitudes is enormous, which is why binary codes other than the 8-4-2-1 type are used industrially. The 8-4-2-1 code is used mainly in computing because of the ease with which arithmetical operations can be carried out on numbers that use this code.

Even if we assume that the groups of 16 bits can be counted out perfectly, what happens if one bit is missed or mistaken? At a frequency of 1 MHz or more it would be hopelessly optimistic to assume that a bit might not be lost or changed. There are tape dropouts and dropins to consider, and discs cannot have perfect surfaces. At such a density of data, faults are inevitable, and some methods must be used to ensure that the groups of 16 bits (**words**) remain correctly gathered together. Whatever method is used must not compromise the rate at which the numbers are transmitted, however, because this is the sampling rate and it must remain fixed. Fortunately, the problems are not new nor unique to audio; they have existed for a long time and been tackled by the designers of computer systems. A look at how these problems are tackled in simple computer systems gives a few clues as to how the designers of audio digital systems went about their task.

Summary

The problem of conversion from analog to digital signals for sound waves is the rate of conversion that is needed. The accepted number of digits per sample is 16, and Shannon's law states that the sampling rate must be at least twice the highest component of the analog signal. For a maximum of 20 kHz, this means a sampling rate of 40 kHz, and for 16-bit signals, this requires a rate of 16×20 thousand bits per second, which is 320,000 bits per second. These bit signals have to be recorded and transmitted in serial form, meaning one after another.

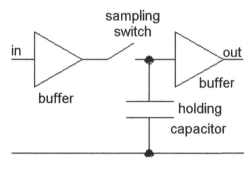

Figure 9.2:

An elementary sample and hold circuit, using a switch to represent the switching actions that would be carried out using metal-oxide-semiconductor field-effect transistors (MOSFETs). When the switch opens, the voltage on the capacitor is a sampled waveform voltage for that instant, and the conversion to digital form of this voltage at the output of the sample and hold circuit must take place before the next sample is taken

Analog to Digital

Converting from an analog into a digital signal involves the quantization steps that have been explained above, but the mechanism needs some explanation in the form of block diagrams. All analog to digital (A/D) conversions start with a sample and hold circuit, illustrated in block form in Figure 9.2.

The input to this circuit is the waveform that is to be converted to digital form, and this is taken through a buffer stage to a switch and a capacitor. While the switch is closed, the voltage across the capacitor will be the waveform voltage; the buffer ensures that the capacitor can be charged and discharged without taking power from the signal source. At the instant when the switch opens, the voltage across the capacitor is the sampled waveform voltage, and this will remain stored until the capacitor discharges. Since the next stage is another buffer, it is easy to ensure that the amount of discharge is negligible. While the switch is open and the voltage is stored, the conversion of this voltage into digital form (quantization) can take place, and this action must be completed before the switch closes again at the end of the sampling period.

In this diagram, a simple mechanical switch has been indicated but in practice this switch action would be carried out using MOSFETs which are part of the conversion IC. To put some figures on the process, at the sampling rate of 44.1 kHz that is used for CDs, the hold period cannot be longer than 22 ms, which looks long enough for conversion − but read on!

The conversion can use a circuit such as is outlined in Figure 9.3, and which very closely resembles the diagram of a digital voltmeter illustrated in Chapter 11. The clock pulses are at a frequency that is much higher than the sampling pulses, and while a voltage is being held at the input, the clock pulses pass through the gate and are counted. The clock pulses are also the input to the integrator, whose output is a rising voltage. When the output voltage from the

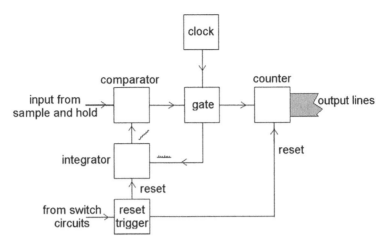

Figure 9.3:
A typical analog to digital (A/D) converter. The gate passes clock pulses that are integrated until the output of the integrator equals the comparator input voltage. The pulse count is the output of the counter, the digital signal. The circuit then resets for the next count

integrator reaches the same level (or slightly above the level) as the input voltage, the gate shuts off, and the counted number at this point is used as the digital signal. The reset pulse (from the sample and hold switch circuit) then resets the counter and the integrator so that a new count can start for the next sampled voltage.

The clock pulse must be at a rate that will permit a full set of pulses to be counted in a sampling interval. For example, if the counter uses 8-bit output, corresponding to a count of 65,538 (which is 2^8), and the sampling time is 20 ms, then it must be possible to count 65,536 clock pulses in 20 ms, giving a clock rate of 3.27 GHz. This is not a rate that would be easy to supply or to work with (though rates of this magnitude are now commonplace in computers), so that the conversion process is not quite as simple as has been suggested here. For CD recording, more advanced A/D conversion methods are used, such as the successive approximation method (in which the input voltage is first compared to the voltage corresponding to the most significant digit, then successively to all the others, so that only eight comparisons are needed for an 8-bit output rather than 65,536). If you are curious about these methods, see the book *Introducing Digital Audio* from PC Publishing.

■ **Note**

The sampling rate for many analog signals can be much lower than the 44.1 kHz that is used for CD recording, and most industrial processes can use much lower rates. ICs for these A/D conversions are therefore mostly of the simpler type, taking a few milliseconds for each conversion. Very high-speed (**flash**) converters are also obtainable that can work at sampling rates of many megahertz.

Digital to Analog

Converting digital to analog is easy enough if a small number of bits are being used and if the rate of conversion is low. Figure 9.4 shows a simple method that uses an operational amplifier for a 4-bit digital signal. If we imagine that the switches represent a set of digital inputs (closed = 1, open = 0) then the digital signals will close or open feedback paths for the operational amplifier, so that the output voltage will be some multiple of the input (reference) voltage) that is set by the total resistance in the feedback path. In practice, of course, the switches are MOSFETs which are switched on or off by the digital inputs on four lines in this example.

Though this system is adequate for 4-bit conversions, it presents impossible problems when you try to use it for 8 or more bits. The snag is that the resistor values become difficult to achieve, and the tolerance of resistance is so tight that manufacturing is expensive. Other methods have therefore been devised, such as current addition and bitstream converters, both beyond the scope of this book. Converters for 16 bits are now readily available at reasonable prices. Flash conversion is dealt with in Chapter 12.

Summary

Conversion between analog and digital signals is straightforward if sampling rates are low and the number of bits is small. More elaborate methods are needed for fast conversion and for 16-bit operation, but the problems have been solved by the IC manufacturers and a large range of conversion ICs can be bought off the shelf.

Serial Transmission

To start with, when computers transmit data serially, the word that is transmitted is not just the group of digits that is used for coding a number. For historic reasons, computers transmit in

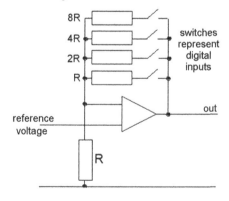

Figure 9.4:
A simple voltage-adding digital to analog (D/A) converter

units of 8-bit bytes, rather than in 16-bit words, but the principles are equally valid. When a byte is transmitted over a serial link, using what is called **asynchronous** methods, it is preceded by one **start bit** and followed by one or two (according to the system that is used) **stop bits**.

Since the use of two stop bits is very common, we will stick to the example of one start bit, eight number bits and two stop bits. The start bit is a 0 and the stop bits are 1s, so that each group of 11 bits that are sent will start with a 0 and end with two 1s. The receiving circuits will place each group of 11 bits into a temporary store and check for these start and stop bits being correct. If they are not, then the digits as they come in are shifted along until the pattern becomes correct. This means that an incorrect bit will cause loss of data, because it may need several attempts to find that the pattern fits again, but it will not result in every byte that follows being incorrect, as would happen if the start and stop bits were not used.

The use of start and stop bits is one very simple method of checking the accuracy of digital transmissions, and it is remarkably successful, but it is just one of a number of methods. In conjunction with the use of start and stop bits, many computer systems also use what is known as **parity**, an old-established method of detecting one-bit errors in text data. In a group of eight bits, only seven are normally used to carry text data (in ASCII code) and the eighth is spare. This redundant bit is made to carry a checking signal, which is of a very simple type.

We can illustrate how it works with an example of what is termed **even parity**. Even parity means that the number of 1s in a group of eight shall always be even. If the number is odd, then there has been an error in transmission and a computer system may be able to make the transmitting equipment try again. When each byte is sent the number of 1s is counted. If this number is even, then the redundant bit is left as a 0, but if the number is odd, then the redundant bit is made a 1, so that the group of eight now contains an even number of 1s. At the receiver, all that is normally done is to check for the number of 1s being even, and no attempt is made to find which bit is at fault if an error is detected. The redundant bit is not used for any purpose other than making the total number even. The process is illustrated in Table 9.2.

Parity, used in this way, is a very simple system indeed, and if two bits in a byte are in error it is possible that the parity could be correct though the transmitted data was not. In addition, the

Table 9.2: Using parity to check that a byte has been received correctly.

Signal byte	0011001	0101111	0101110	1110010
Even parity added	1	1	0	0
Signal sent	10011001	10101111	00101110	01110010
Received byte	10001001	10101111	10101000	01110010
Parity check	odd	even	odd	even
Result	error	OK	error	OK

The numbers each use seven bits, and one extra parity bit is added, in this case to make the count of 1s an even number. On reception, the byte can be checked to find if its parity is even. The parity bit is located on the left hand side.

parity bit itself might be the one that was affected by the error so that the data is signaled as being faulty even though it is perfect. Nevertheless, parity, like the use of start bits and stop bits, works remarkably well and allows large masses of computer text data to be transmitted over serial lines at reasonably fast rates. What is a reasonably fast rate for a computer is not, however, very brilliant for audio, and even for the less demanding types of computing purposes the use of parity is not really good enough and much better methods have been devised. Parity has now almost vanished as a main checking method because it is now unusual to send plain (7-bit) text; we use formatted text using all 8 bits in each byte, and we also send coded pictures and sound using all 8 bits of each byte, so that an added parity bit would make a 9-bit unit (not impossible but awkward, and better methods are available).

The rates of sending bits serially over telephone lines in the pre-broadband days ranged from the painfully slow 110 bits per second (used at one time for teleprinters) to the more tolerable 56,000 bits per second. Even this fast rate is very slow by the standards that we have been talking about, so it is obvious that something rather better is needed for audio information. Using direct cable connections, rates of several million bits per second can be achieved, and these rates are used for the universal serial bus (USB) system that is featured in modern computers. We will look at broadband methods later.

As a further complication, recording methods do not cope well with signals that are composed of long strings of 1s or 0s; this is equivalent to trying to record square waves in an analog system. The way round this is to use a signal of more than 8 bits for a byte, and using a form of conversion table for bytes that contain long sequences of 1s or 0s. A very popular format that can be used is called eleven-to-fourteen (ETF), and as the name suggests, this converts each 11-bit piece of code (8 bits of data plus 3 bits used for error checking) into 14-bit pieces which will not contain any long runs of 1s or 0s and which are also free of sequences that alternate too quickly, such as 01010101010101.

All in all, then, the advantages that digital coding of audio signals can deliver are not obtained easily, whether we work with tape or with disc. The rate of transmission of data is enormous, as is the bandwidth required, and the error-detecting methods must be very much better and work very much more quickly than is needed for the familiar personal computers that are used to such a large extent today. That the whole business should have been solved so satisfactorily as to permit mass production is very satisfying, and even more satisfying is the point that there is just one worldwide CD standard, not the furiously competing systems that made video recording such a problem for the consumer in the early days.

For coding television signals, the same principles apply, but we do not attempt to convert the analog television signal into digital because we need only the video portion, not the synchronization pulses. Each position on the screen is represented by a binary number which carries the information on brightness and color. Pictures of the quality we are used to can be obtained using only 8 bits. Once again, the problems relate to the speed at which the information is sampled, and the methods used for digital television video signals are quite unlike those used for the older system, though the signals have to be converted to analog form before they are applied to the guns of a cathode-ray tube (CRT). Even this conversion becomes unnecessary when CRTs are replaced by color liquid crystal display (LCD) screens, because the IC that deals with television processing feeds to a digital processor that drives the LCD dots.

A more detailed description and block diagram of CD replay systems is contained in Chapter 12.

Summary

Digital coding has the enormous advantage that various methods can be used to check that a signal has not been changed during storage or transmission. The simplest system uses parity, adding one extra bit to check that number of 1s in a byte. More elaborate systems can allow each bit to be checked, so that circuits at the far end can correct errors. In addition, using coding systems such as ETF can avoid sequences of bits that are difficult to transmit or record with perfect precision.

CHAPTER 10

Gating and Logic Circuits

Gates

Digital circuits come in several varieties, but one very basic and important type is the **gate**, which is used for controlling actions rather than just for counting. More than 100 years before digital electronics was established, an English mathematician called George Boole proved that all of the statements in human logic could be expressed by combinations of three rules which he called AND, OR, and NOT. This logic system is now known as **Boolean logic**. We will see later how these gates can also be used for simple arithmetic actions.

Definition

A gate in digital electronics means a circuit whose output is a 1 only for some specified combination of inputs. This type of circuit is sometimes referred to as a **combinational circuit**.

■ Note

Do not confuse this with an **analog gate circuit**, which switches an analog waveform on or off depending on the state of a gating input signal.

■

The importance of all this is that, if we can provide digital gate actions corresponding to these three rules of AND, OR, and NOT, we can construct a circuit that will give a 1 output for any set of logical rules. For example, if we want to have an electric motor switched on when a cover is down, a switch is up, and a timer has reached zero, or when an override switch is pressed, then this set of rules can be expressed in terms of AND, OR, and NOT, and a gate or set of gates can carry out the action.

Figure 10.1 shows the symbols that are used for the three main gate types, the AND, OR, and NOT gates, using two-input gates in the example. The action of these gates will be discussed in detail shortly, but for the moment note that the small circle shown in the NOT symbol is used to mean inversion, converting 1 to 0 or 0 to 1. We can combine the other basic symbols with the NOT circle symbol to give symbols for other gate actions. The early digital integrated circuits (ICs) would typically contain four gates of one type per chip, but modern electronic equipment

Electronics Simplified. DOI: 10.1016/B978-0-08-097063-9.10010-X

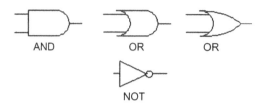

Figure 10.1:
Gate symbols for the basic actions of AND, OR, and NOT. These are the internationally used symbols and this drawing shows two variants on the OR symbol

is more likely to use custom-made chips, with all the gates and their connections formed in one process, using thousands or millions of gates on one chip.

Once again, this makes it more useful to show block diagrams rather than gate circuit details. A gate circuit diagram will consist of a large number of gate symbols with joining lines so that the output of a gate will be connected to one or more other gate inputs. Provided that we know what each basic type of gate does, we can analyze the action of complete gate circuits. In this book we are concerned more with block diagrams than with gate circuits, but some knowledge of gate circuits is useful, and in any case, these are closer in spirit to block diagrams than to circuit diagrams.

Summary

Logic circuits exist to carry out a set of logic actions such as are used for controls for washing machines, tape-recorder drives, computer disk drives, security systems, and a host of industrial control actions. Simple arithmetic actions can also be carried out using logic circuits. All logic actions, however complicated, can be analyzed into simple actions that are called AND, OR, and NOT, so that circuits, called gates, which carry out such actions, are the basis of logical circuits.

■ Note

Remember that ICs are classed as medium-scale integration (MSI), large-scale integration (LSI), very large-scale integration (VLSI), etc., by the number of equivalent gate circuits on a single chip.

■

Definition

The action of any gate can be expressed in a **truth table**. This is just a table that shows all the possible inputs to the gate, and the output for each set of inputs. Remember that each input can be 0 or 1 only, so that each input contributes two possible outputs. The total number of outputs is equal to 2^n, where n is the number of inputs.

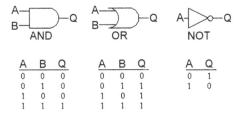

Figure 10.2:
Truth tables for gates. The AND and OR gates are illustrated in two-input form

For example, if there are four inputs to a gate, then the number of possibilities is $2^4 = 16$, and its truth table will consist of 16 lines. For a lot of truth tables, there is only one output that is different from the rest, and it is easier to remember which one this is than to try to remember the whole of a truth table.

■ Note

Truth tables are the simplest way of showing what a small-scale gate or gate circuit does, but they are impossibly clumsy when we try to use them on complex gates with a large number of inputs. The more useful method for professional use is called **Boolean algebra** but, like other mathematical methods, it is beyond the scope of this book. There are computer programs that will analyze the action of a gate circuit using Boolean algebra.

■

Figure 10.2 shows truth tables for the basic two-input AND, OR, and NOT gates. Of these, the NOT gate is a simple one, with just one input and one output. Its action is that of a logic inverter. If the input is 0, then the output is not0, which is 1. If the input is 1, then the output is not1, which is 0. The other two gate types permit more than one input, and the examples show two inputs, the most common number. The action of the AND gate is to give a 1 output only when both inputs are at 1, and a 0 output for any other combination. The action of the OR gate is to give a 0 output when each input is 0, but a 1 for any other combination of inputs. The same arguments apply to gates with more than two inputs.

Summary

A truth table is a simple but clumsy way of showing what the output of a gate or gate circuit will be for each and every possible combination of inputs. The alternative is to use Boolean algebra, a process that is greatly simplified by computer programs that carry out an analysis of gate circuits.

NAND and NOR Gates

Two particularly useful gate types can be made by combining the action of an inverter with that of the AND and OR gates. The combination of NOT and AND gives the NAND gate, whose

NAND gate

A	B	Q
0	0	1
0	1	1
1	0	1
1	1	0

Figure 10.3:
The NAND gate, with symbol, truth table and equivalent gate circuit

symbols and truth table (for two inputs) are shown in Figure 10.3. The action of this gate is that the output is 0 only when all of its inputs are at 1, which is the action of the AND gate followed by an inverter. The combined action of the OR gate and a following inverter gives the NOR gate, whose symbols and truth table are shown in Figure 10.4. The output of this gate will be at logic 0 when any one (or more) of its inputs is at logic 1.

There is one further gate that is often used and which is called **exclusive-OR** (XOR). This action (Figure 10.5) is closer to what we normally mean by the word 'or' (meaning one or the other but not both), and the output is 1 if either input is 1, but not when both inputs are zero or both are 1. The diagram also shows that the XOR gate is equivalent to the action of a circuit made using an OR, AND, and NAND gate combination.

NOR gate

A	B	Q
0	0	1
0	1	0
1	0	0
1	1	0

Figure 10.4:
The NOR gate with symbol, truth table and equivalent gate circuit

XOR gate

A	B	Q
0	0	1
0	1	0
1	0	0
1	1	1

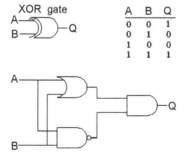

Figure 10.5:
The XOR gate with truth table and equivalent gate circuit

■ Note

If you would like to read further about gates and other logic circuits, with details of the more advanced methods such as Boolean algebra, take a look at the book *Digital Logic Gates and Flip-Flops*, from PC Publishing.

■

Summary

A truth table is a simple way of expressing the action of a logic circuit, and the standard gates called OR, AND, and NOT can all be illustrated in this way. Gates in IC form often consist of the NAND and NOR type of gates, equivalent to an AND or OR, respectively, followed by NOT. These inverting gates are easier to produce and the action is often more useful than that of the simpler AND or OR type. The XOR gate is another useful type which gives an action closer to the normal meaning of OR, as 'one or the other but not both'.

Analyzing Gate Systems

A circuit that has been made up by connecting several standard gates together, which has several inputs and an output, can be analyzed to find what its action is. This analysis can be done by drawing up truth tables, or by a method called Boolean algebra. The truth-table method is simpler, but more tedious than the Boolean algebra method, which is not dealt with in this book. The method of analysis by truth table can be summarized in a few rules.

1. Letter each input to the circuit (A, B, C) and also each point where the output of one gate is connected to the input of another gate, using different letters for each point. Label the final output as Q.
2. Draw up a blank truth table, using one column for each letter that has been allocated, and with 2^n rows, where n is the number of signal inputs to the circuit.
3. Write in every possible combination of inputs. This is most easily done by starting with 0000 and continuing in the form of a binary count (0001, 0010, 0011, 0100, 0101) up to an input which consists entirely of 1s.
4. Knowing the truth tables for the standard gates, write in the logic states (0 or 1) for the outputs of the gates at the inputs in each line of the truth table.
5. The first set of outputs will now be the inputs for the next set of gates, so that their outputs can be written into the truth table.
6. Continue in this way until the truth table has been completed.

As an example, Figure 10.6 shows a logic diagram for an electronic combination lock. This is a simple design, with four main inputs, and therefore 16 combinations, ignoring the unlock input E. The lock is arranged so that only the correct combination of inputs will open the

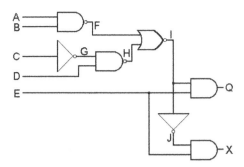

Figure 10.6:
A gate circuit used as an example for analysis using truth tables

lock, and any other combination will cause an alarm to sound, so that it cannot be solved by trying each possible combination.

The inputs, A, B, C, and D are from switches which are to be set in the pattern needed to open the lock. When these switches have been set, pressing the button E will cause the door to unlock (Q = 1) if the combination is correct, or cause the alarm to sound (X = 1) if the combination is incorrect. To analyze this digital circuit, label the inputs as shown in Figure 10.7.

The important inputs are A, B, C, and D, because E is used only after all the others have been set into the correct pattern. The intermediate points, where the output of one gate drives the input of another gate, can now be labeled F, G, H, I, and J as shown in Figure 10.7. Because there are four main inputs, 16 lines of truth table will be needed.

There will be one column for each letter which has been used, but the column for the E input can be placed next to the Q column because the E input is used only when the Q output is decided (after all the other inputs have been set). The logic voltage of E can be written as 1 in each row because the lock will act only when E is set to 1 (the activating button is pushed). All of the possible A, B, C, D inputs can now be written down, starting with 0000, and going through a binary count to 1111, a total of 16 rows in the truth table of Figure 10.7.

We can now analyze the circuit (Figure 10.8). Inputs A and B are inputs of a NAND gate. whose logic is that the output is 0 only when both inputs are 1. The F column, which is the output of this gate, therefore has a 0 entered for the last four rows of the table, when both A and B are at logic 1, and a 1 entered for all other rows The G column is just the inverse of the C column, so that its values can now be written in.

The values in the H column are the outputs of another NAND gate whose inputs are G and D, so that the output is 0 only when G = 1 and D = 1, as shown. The values in columns F and H are now the inputs to a NOR gate whose outputs are written into column I.

A	B	C	D	F	G	H	I	J	E	Q	X
0	0	0	0						1		
0	0	0	1						1		
0	0	1	0						1		
0	0	1	1						1		
0	1	0	0						1		
0	1	0	1						1		
0	1	1	0						1		
0	1	1	1						1		
1	0	0	0						1		
1	0	0	1						1		
1	0	1	0						1		
1	0	1	1						1		
1	1	0	0						1		
1	1	0	1						1		
1	1	1	0						1		
1	1	1	1						1		

Figure 10.7:
The table with inputs filled in and intermediate values provided for

The logic of the NOR gate is that the output is 1 only when both outputs are at 0, and this occurs only on one line of the table, when $A = 1$, $B = 1$, $C = 0$, $D = 1$. When $I = 1$ and $E = 1$, the output of the AND gate then gives $Q = 1$, so that the lock opens. For any other combination of inputs at A, B, C, and D, the value of Q is 0 and the value in J is 1 (because of the inverter) and the combination of $E = 1$, $J = 1$ causes $X = 1$, sounding the alarm but keeping the door locked. In addition, pressing switch E before any of the others are set will also cause the alarm to sound. The action of this set of gates is to open the lock only for the correct combination of inputs and to sound the alarm for an incorrect combination. Figure 10.8 shows the final state of the truth table.

Summary

Any gate circuit can be analyzed by drawing up a truth table, or a set of truth tables. Though this can be tedious when a circuit has a large number of inputs, it is simpler than the Boolean algebra alternative method (though this can be carried out using a computer application). The method relies on using lettering to identify all inputs, outputs, and intermediate points, and drawing up the truth table in stages, starting with all possible combinations at the inputs.

A	B	C	D	F	G	H	I	J	E	Q	X
0	0	0	0	1	1	1	0	1	1	0	1
0	0	0	1	1	1	0	0	1	1	0	1
0	0	1	0	1	0	1	0	1	1	0	1
0	0	1	1	1	0	1	0	1	1	0	1
0	1	0	0	1	1	1	0	1	1	0	1
0	1	0	1	1	1	0	0	1	1	0	1
0	1	1	0	1	0	1	0	1	1	0	1
0	1	1	1	1	0	1	0	1	1	0	1
1	0	0	0	1	1	1	0	1	1	0	1
1	0	0	1	1	1	0	0	1	1	0	1
1	0	1	0	1	0	1	0	1	1	0	1
1	0	1	1	1	0	1	0	1	1	0	1
1	1	0	0	1	1	1	0	1	1	0	1
1	1	0	1	0	1	0	1	0	1	1	0
1	1	1	0	0	0	1	0	1	1	0	1
1	1	1	1	0	0	1	0	1	1	0	1

Figure 10.8:
The table completed, using the truth tables for the gates to fill in the intermediate values

Arithmetic Circuits

Definition

An arithmetic circuit is a set of gates with a separate set of inputs for each number that has to be processed. The gates are connected so as to carry out an arithmetic action and the outputs of the gate circuit are the digits of the result (addition, subtraction, multiplication, or division).

Adder Circuits

Definition

The rules of binary addition for two bits are:

$0 + 0 = 0$
$0 + 1 = 1$
$1 + 0 = 1$
$1 + 1 = 0$ and carry 1 (the binary number 10; denary 2).

The simplest possible adder circuit for binary digits is called a **half-adder**, and it allows two bits to be added, with a main output and a carry bit. The truth table is illustrated in Figure 10.9, and this illustration also indicates that the half-adder can be constructed by using a combination of an exclusive-OR gate and an AND gate. The carry bit is 0 except when both inputs bits are 1, which is as required by the rules of binary arithmetic.

This half-adder is such a useful circuit that it is made in IC form in its own right, and it can, in turn, be used to create other circuits. The name 'half-adder' arises because it can be used only as a first stage in an adder circuit. If we need to add only two bits, the half-adder is sufficient, but if we need to add, say, eight pairs of bits, as when we add two bytes, then the other adders will have three inputs: the bits that are to be added plus the carry bit from the previous stage. For example, adding 1011 and 0011 can use a half-adder for the lowest order pair, giving a 0 output and a carry bit. The addition of the next bits is $1 + 1 + 1$, because of the carry, and

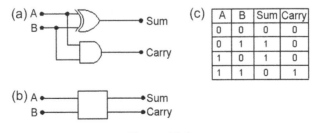

Figure 10.9:
(a) A half-adder circuit, with (b) symbol and (c) truth table

this gives a 1 output and a 1 carry. The next addition uses the carry from the previous stage, and the last addition uses no carry:

Carry	0	1	1	0
A	1	0	1	1
B	0	0	1	1
Sum	1	1	1	0

If we have the half-adder in IC form, we can connect two half-adders along with an OR gate to provide a full adder, which allows a carry input as well as a carry output. Figure 10.10 shows the circuit, its truth table, and the block representation.

■ Note

A full adder can be used as a half-adder if the carry input is connected permanently to the 0 voltage level.

■

In practice, adders can be serial or parallel. A serial adder works on each pair of bits (and any carry) at a time, adding the figures much as we do with pencil and paper. The only complication is the need to store the bits, and we shall see how this is done in Chapter 11. A parallel adder will have as many inputs as there are bits to add, and the outputs will appear as soon as the inputs are present. This is very much faster, and though the circuit is more complicated, this is no problem when it can be obtained in IC form.

Figure 10.11 shows what is involved in a typical parallel adder circuit for two 4-bit binary numbers. In this circuit, all of the inputs are applied at the same time, and the only delay is due to the carry bits, because the addition at each of the later stages is not complete until the carry bit has been generated at the previous stage. This typically takes a few nanoseconds per stage, and the process of passing the carry bit from stage to stage is called **rippling through**.

■ Note

Since there is never a carry input at the least significant stage, this could use a half-adder, but when the circuit is constructed in IC form a full adder with the first carry set to zero is normally used.

■

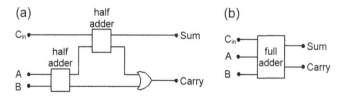

Figure 10.10:
(a) The full adder constructed from gates and (b) its symbol

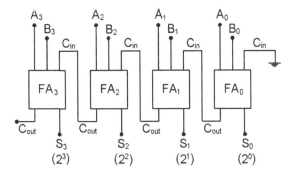

Figure 10.11:
A 4-bit parallel adder for two binary numbers

Subtraction

Definition

The rules for binary subtraction are:

$0 - 0 = 0$
$0 - 1 = -1$ (borrow 1)
$1 - 0 = 1$
$1 - 1 = 0$

where the -1 entry represents a borrow (or carry-back) action from the next higher bit, rather than the carry-forward operation that is used for addition.

Figure 10.12 shows how a half-subtractor circuit can be constructed using a NOT gate added to the simple half-adder circuit. Once again, the 'half' reminds you that this circuit does not provide for a borrow, and a two-bit full subtractor circuit can be made using half-subtractors as shown in Figure 10.13. A single stage like this can be used for serial subtraction.

As you might expect, a full subtractor can be made in parallel form, and an example which provides for two 4-bit numbers is illustrated in Figure 10.14 using full subtractor units. This, however, can also be built using full adder circuits, using a method of subtraction that we need not worry about at the moment (twos complement with end-around carry). The point is that it

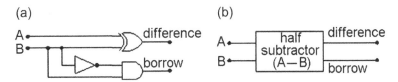

Figure 10.12:
(a) The half-subtractor in gate form and (b) as a symbol

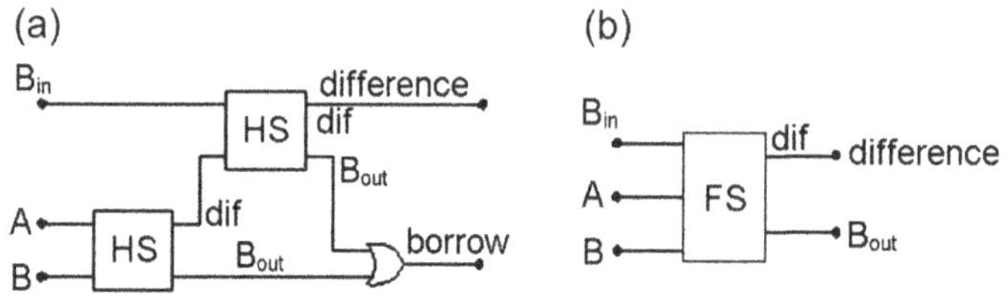

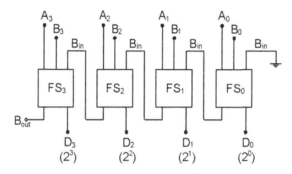

Figure 10.13:
(a) The full subtractor in gate form and (b) as a symbol

Figure 10.14:
A 4-bit full parallel subtractor

allows subtractors to be made using the same IC components as adders, making production more economical.

Multiplication and Division

Definition

The rules of binary multiplication are:

$0 \times 0 = 0$
$0 \times 1 = 0$
$1 \times 0 = 0$
$1 \times 1 = 1$

and the rules for division are:

$0 \div 1 = 0$
$1 \div 1 = 1$

There are only two division rules, because division by zero is meaningless.

Multiplication and division cannot be carried out by gates alone, because they must be done stage by stage. Consider, for example, a multiplication carried out in the same way as we would multiply two denary numbers:

```
    1101
    1010
    0000
    1101.
    0000..
    1101...
 10000010
```

This starts by multiplying each bit in the upper number by the lowest order bit in the second number, producing in this example a set of zeros. Each bit in the second number (the multiplier) will result in a set of 4 bits (in this example) for the result, and each successive set is shifted left by one place. This action of storage and shifting requires the use of **registers** (see Chapter 11), and this is followed by an addition, in this case of four 4-bit numbers, to find the final result, which may contain a carry to an extra bit position.

Division requires the same types of actions and follows the methods used for long division of denary numbers (once familiar to eight-year-olds, but now a lost art in schools). Just as well we can assemble arithmetic circuits that are ready-made for us.

Summary

Arithmetic can be carried out using gates. The simpler addition and subtraction actions can be carried out using only gates, but the actions of multiplication and subtraction require storage and shifting and are dealt with using **registers**. These actions can be used in serial form (repeating the action for each position in a binary number) or in parallel form (with as many arithmetic units as there are bits in the binary numbers).

Code Changing

The need to use codes for digital information often leads to problems in transmitting the digital signals, because we very often cannot provide a signal line for each code signal. Suppose, for example, that we have a set of eight switches, any one of which can be ON (1) or OFF (0), but not allowing more than one switch to be ON at a time. Suppose also that we need to pass this information to some unit which might be a controller or some type of recorder. One obvious method is to use eight lines, one for each switch. A less obvious method is to use the binary numbers for 0 to 7 (a total of eight codes), which cover the range 000 to 111, using only three lines. A circuit that carries out this change from using eight lines to using three lines is called an 8 to 3 **encoder**. Encoders are used to reduce the number of signal channels needed for data.

■ Note

The word **multiplexer** is often used interchangeably with **encoder**, though strictly speaking the two are not identical. The opposite actions are carried out by the **decoder** or **demultiplexer**.

■

Definition

An encoder is used to convert a set of 2^n signal lines, on which only one line at a time can be at logic 1, into binary signals on n lines. A decoder performs the conversion from n lines of binary signals to 2^n lines, of which only one at a time can be at logic 1. A multiplexer (**Mux**) is used to select one signal from signals on n input lines and produce the selected signal as an output on one line. A demultiplexer (**Demux**) has a single input which varies with time, and will provide outputs on n lines.

Figure 10.15 shows an encoder block and truth table for the example of eight to three transformation. A decoder for three lines to eight, also shown here, would work in the opposite way, with a 3-bit binary input and a set of eight lines out, only one of which could be at level 1 for each combination of input bits. You might, for example, use the outputs from the decoder to operate warning lights, if only one at a time needed to be illuminated.

Definition

For any decoder, the binary number input is often referred to as an **address number**. This name crops up again when we look at computer circuits.

An encoder is needed when each of a set of inputs must generate a unique code pattern. For example, imagine a set of 16 keys (a **hexadecimal keypad**) for which each of the 16 input key characters must generate a unique 4-bit output code; we assume that only one key is pressed at a time. This encoder would require 16 inputs, only one of which is selected at any one time, and would provide four output lines to provide the parallel bit pattern. If we needed to reconstitute

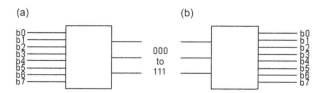

(a)   (b)

b0   000   b0
b1   to   b1
b2   111   b2
b3   b3
b4   b4
b5   b5
b6   b6
b7   b7

Figure 10.15:
(a) Encoder and (b) decoder in block form. The illustration shows 8 to 3 encoding and 3 to 8 decoding

the pattern of lines at the input, we would use a decoder that would have four input lines and would use the bits on these lines to select one and only one of 16 output characters.

Encoders and decoders can be made using gates, and Figure 10.16 shows an example of a 2 to 4 line decoder using inverters and AND gates, along with a truth-table analysis of the action.

Now let's look at a multiplexer. Figure 10.17 shows in block form what a 4 to 1 multiplexer is expected to do. There are four data inputs, two address inputs, and a strobe input. The address inputs are used to select one of the four data lines, using an encoding action (which is why encoders and multiplexers are often taken to be almost identical). The data on the selected line is transferred to the single output line when the signal on the strobe line is low (logic 0). This allows all of the data to be sent down one line, a bit at a time, by establishing an address number, pulsing the strobe input, and repeating this for another address number until you have cycled through all the address numbers.

Definition

A strobe pulse is one that is applied at a set time to allow an action to be carried out. The name was originally used in radar to mean a pulse that was used to enable the receiver for a time that excluded return pulses from nearby targets.

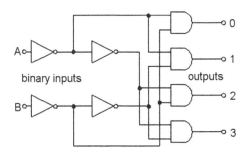

Figure 10.16:
A decoder in gate form for 2 to 4 decoding

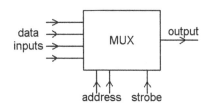

Figure 10.17:
A multiplexer in block form, in this example a 4 to 1 Mux with two address lines and a strobe line

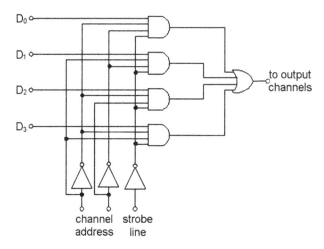

Figure 10.18:
A 4 to 1 multiplexer in gate form

■ Note

This type of action, allowing a mass of data to be selected for transmission down a single line, is called **time-division multiplexing**. The use of a carrier modulated in more than one way (such as the color television sub-carrier) is another form of multiplexing. Multiplexing of the time-division type is very important for digital television and radio, enabling several programs to share a single bandwidth channel.

■

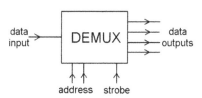

Address		Data		Strobe		Outputs	
A	B			Y_0	Y_1	Y_2	Y_3
0	0	0	0	0	1	1	1
0	1	0	0	1	0	1	1
1	0	0	0	1	1	0	1
1	1	0	0	1	1	1	0

Figure 10.19:
Block for a 1 to 4 demultiplexer

Figure 10.18 shows a gate circuit that can carry out the action of multiplexing a set of four inputs. For four inputs, a 2-bit address number is needed to select which input will be connected to the output line when the strobe input goes to logic 0. The opposite action of a demultiplexer is shown in block form in Figure 10.19. There is a single data input and a strobe input, along with the address inputs, in this case using two lines for the addresses of four output lines. Once again, the data on the input line will appear at the selected output line when the strobe pulse voltage is low, and the truth table is in Figure 10.19 with the selected output data line in this example taking the logic value of 0 (the same as the data line input) when the strobe appears (assuming that strobing occurs when the strobe input is low).

Summary

Encoders, decoders, multiplexers, and demultiplexers are used to change the way that digital codes are carried. Usually, one of the codes will be 8-4-2-1 binary. The distinguishing feature of the multiplexer is that it allows any number of input lines to be used to place signals on a single output line, controlled by a selection address number and an enabling strobe pulse. The demultiplexer carries out the opposite action, assuming that the strobe pulses are correctly synchronized.

CHAPTER 11

Counting and Correcting

Sequential Circuits

Gate circuits are one basic type of digital circuit, the **combinational** circuit. The other basic type of digital circuit is the **sequential** type. The output of a sequential circuit, which may mean several different output signals on separate terminals, will depend on the **sequence** of inputs to the circuit. One simple example is a counter integrated circuit (IC), in which the state of the outputs will depend on how many pulses have arrived at a single input. Compare this with the combinational (gate) circuit in which there are several inputs and the output depends on the combination of inputs.

The basis of all sequential circuits is a circuit called a **flip-flop**, and the simplest flip-flop circuit is called the S-R flip-flop, with the letters S and R meaning set and reset. Though this type of flip-flop is not manufactured in IC form (because it is just as simple to manufacture more complex and more useful types), it illustrates the principle of sequential circuits more clearly than the more complicated types.

Definition

A flip-flop is a circuit whose output(s) change state for some sequence of inputs and which will remain unchanged until another sequence of inputs is used. Unlike a gate, simply changing the inputs to a sequential circuit does not necessarily change the outputs.

■ **Note**

One feature of flip-flops is the use of two outputs, conventionally labeled as $\bar{Q}$ and Q# (or $\overline{Q}$). The Q output is the main output, and the Q# output is the inverse of the Q output.

■

Figure 11.1 shows the block diagram for one variety of simple S-R flip-flop along with a table of inputs and outputs. The notable point is that the table shows five lines, and that the line that uses 0 for both inputs appears twice. The output for $S = 0$ and $R = 0$ can be either $Q = 1$ or $Q = 0$, and the value of Q depends on the **sequence** that leads to the $S = 0$, $R = 0$ state. If you use inputs $S = 1$, $R = 0$, then $Q = 1$, and this value will remain at 1 when the input changes to $S = 0$ and $R = 0$. Similarly, if the inputs are $S = 0$, $R = 1$, the Q output becomes 0, and this remains at 0 when the inputs change to $S = 0$, $R = 0$.

Electronics Simplified. DOI: 10.1016/B978-0-08-097063-9.10011-1

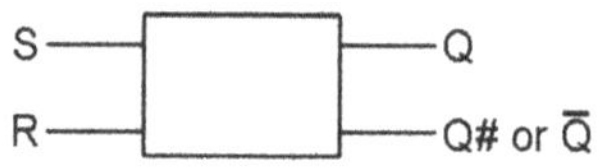

S ────┌──────┐──── Q
R ────└──────┘──── Q# or $\bar{Q}$

S	R	Q	Q#	
1	1	0	0	forbidden state
1	0	1	0	set
0	0	1	0	store
0	1	0	1	reset
0	0	0	1	store

Figure 11.1:
The S-R (or R-S) flip-flop with its state table

Note

One state, with $S=1$ and $R=1$, is forbidden because all the uses for flip-flops require Q# always to be the opposite of Q, and the S-R flip-flop violates this requirement when $S=1$ and $R=1$.

The meanings of S and R are now easier to understand. Using $S=1$ and $R=0$ will **set** the output, meaning that Q changes to 1. Using $S=0$, $R=1$ will **reset** the output, meaning that Q changes to 0. The input $S=0$, $R=0$ is a storing (or **latching**) input, making the flip-flop remain with the output that it had previously. Though the single S-R flip-flop is not manufactured in IC form, it is useful (as a latch, see later) and can be constructed using two NAND gates as shown in Figure 11.2. Multiple S-R latch chips are available.

Summary

Sequential circuits are the other important digital type, used in counting and for memory actions. The simplest type is the S-R flip-flop (or latch), whose output(s) can be set by one pair of inputs and reset by reversing each input. The output is held unchanged when both input signals are zero. Sequential circuits can be created using gates, emphasizing the importance of the gate as a fundamental digital circuit.

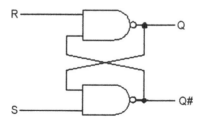

Figure 11.2:
How an S-R flip-flop can be constructed using NAND gates. Another version can be made using NOR gates, with a slightly different state table

The J-K Flip-Flop

Though simpler types exist, the most important type of flip-flop is that known as the master–slave J-K flip-flop, abbreviated to J-K. This is a **clocked circuit**, meaning that the action of the IC is carried out only when a pulse is applied to an input labeled **clock**. This is the type of flip-flop that is most commonly manufactured in IC form because it can be used in place of so many other flip-flops.

Using this type of flip-flop allows the actions of a number of such circuits to be perfectly synchronized, and it also avoids the kind of problems that can arise in some types of gate circuits when pulses arrive at slightly different times. These problems are called **race hazards**, and their effect can be to cause erratic behavior when a circuit is operated at high speeds. When clocking is used, the circuits can be operated **synchronously**, meaning that each change takes place at the time of the clock pulse, and there should be no race hazards.

The J-K flip-flop uses three main inputs, labeled J, K, and clock (Ck). Of these, the J and the K are **programming inputs** whose voltage levels will control the action of the flip-flop at the time when the clock pulse triggers the circuit. Because there are two programming inputs, and each of them can be at either of two levels (0 or 1), there are four possible modes of operation of the J-K flip-flop.

We can describe each of those modes of action in terms of what voltage changes occur at the output when the clock-pulse edge arrives. The usual **triggering edge** is the trailing back edge, the level 1 to level 0 transition of a pulse, and in Table 11.1, the condition of the Q output before the arrival of trailing edge is indicated as Qn and the condition after the edge as Q_{n+1}.

The input for this table is the clock pulse, with the J and K terminals used for programming, and outputs are taken from the Q and Q# terminals. Remember that the Q# output is *always* the inverse of the Q output. In older texts this is often shown by

Table 11.1: The condition of the Q output before the arrival of the trailing edge is indicated as Qn and the condition after the edge as Q_{n+1}.

J	K	Qn	Q_{n+1}	Comment
0	0	0	0	No change, latched
0	0	1	1	No change, latched
0	1	0	0	Reset on clock
0	1	1	0	Reset on clock
1	0	0	1	Set on clock
1	0	1	1	Set on clock
1	1	0	1	Toggle
1	1	1	0	Toggle

drawing a bar over the Q. The table shows the possible states of inputs before and after the clock pulse, showing how the voltages on the J and K pins will determine what the outputs will be after each clock pulse.

The four modes which are indicated in Table 11.1 are as follows.

- The **hold** mode. With J = 0 and K = 0, the output is unchanged by clock pulses. This avoids the need for further gating if you want to keep the flip-flop isolated.
- The **set** mode, with J = 1, K = 0. Whatever the state of the outputs are before the clock pulse edge, Q will be set (to 1) at the triggering edge and will stay that way.
- The **reset** mode, with J = 0, K = 1. Whatever the state of the outputs before the triggering edge, Q will be reset (to 0) at the edge and will stay that way.
- The **toggle** mode, with J = 1, K = 1. The output will change over each time the circuit is clocked.

In addition to the inputs J and K, which affect what happens when a clock pulse comes along, these flip-flops are usually equipped with S and R terminals, which are also called preset (Pr) and reset (R) terminals. These will affect the Q and Q# outputs of the flip-flop immediately, irrespective of the clock pulses. For example, applying a zero input to the R input will reset the flip-flop whether there is a clock pulse acting or not. This action is particularly useful for counters that have to be reset so as to start a new count.

Summary

The J-K flip-flop uses two programming inputs labeled J and K, whose voltages determine how the flip-flop will operate when a clock pulse is applied. One particularly useful input is J = 1, K = 1, which allows the flip-flop to reverse (or **toggle**) the output state at each clock pulse, so that its output is at half the pulse rate of the clock. Two inputs labeled S and R (or Pr and R) allow the output to be set (preset) or reset independently of the clock pulses.

The use of two programming inputs, J and K, therefore causes a much more extensive range of actions to be available. Note that if both the J and K terminals are kept at level 1, the flip-flop will toggle so that the output changes over at each clock pulse, as is required for a simple binary counter (Figure 11.3).

Flip-flops are the basis of all counter circuits, because the toggling flip-flop is a single-stage scale-of-two counter, giving a complete pulse at an output for each two pulses in at the clock terminal. By connecting another identical toggling flip-flop so that the output of the first flip-flop is used as the clock pulse of the second, a two-stage counter is created, so that the voltages at the Q outputs follow the binary count from 0 to 3, as Figure 11.4 shows. This principle can be extended to as many stages as is needed, and extended counters of this type can be used as timers, counting down a clock pulse which can initially be at a high frequency.

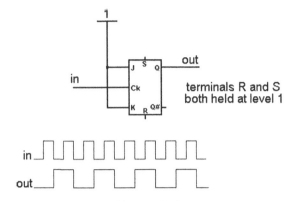

Figure 11.3:
Using a J-K flip-flop as a counter or frequency divider

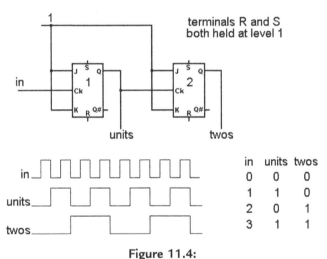

Figure 11.4:
A two-stage simple counter using J-K flip-flops

The type of counter that uses toggling flip-flops in this way is called **asynchronous**, because the clock inputs to the flip-flops do not occur at the same time and the last flip-flop in a chain like this cannot be clocked until each other flip-flop in the chain has changed. For some purposes, this is acceptable, but for many other purposes it is essential to avoid these delays by using **synchronous** counter circuits in which the same input clock pulse is applied to all of the flip-flops in the chain, and correct counting is assured by connecting the J and K terminals through gates. Circuits of this type are beyond the scope of this book, and details can be found in any good text of digital circuit techniques. In this book we shall, as usual, concentrate on block diagrams which are not concerned with whether a counter is synchronous or not.

Summary

The toggling flip-flop is the basis of all counter circuits, which use the clock as an input and the Q outputs as the outputs. In such a circuit, the Q outputs provide a binary number which represents the number of complete clock pulses since the counter was reset.

Counter Uses

Apart from their obvious applications to clocks and watches, counters are used extensively in electronics, and particularly in modern measuring instruments. ICs exist to provide for measurement of all the common quantities such as direct current (DC) voltage, signal peak and root mean square (RMS) amplitude, and frequency, using counting methods. Most of the modern ICs of this type provide for digital representation, so that the outputs are in a form suitable for feeding to digital displays. The simplest application of counters to instruments is for frequency measurement.

Frequency Meter

A frequency meter makes use of a high-stability oscillator, normally crystal controlled, to provide clock pulses. In its simplest form, the block diagram is as shown in Figure 11.5, with the unknown frequency used to open a gate for the clock pulses. The number of clock pulses passing through the gate in the time of one cycle of the input will provide a measure of the input frequency as a fraction of the clock rate. For example, if the clock rate is 10 MHz and 25 pulses of the unknown frequency pass the gate, then the input frequency is 10/25 MHz, which is 400 kHz. A counter circuit can find the number and display it in terms of the clock frequency to give a direct reading of the input frequency.

In this simple form, the frequency meter could not cope with input frequencies that are greater than the clock rate, nor with input frequencies that would be irregular sub-multiples. For example, it cannot cope with 3.7 gated pulses in the time of a clock cycle. Both of these problems can be solved by more advanced designs. The problem of high frequencies can be

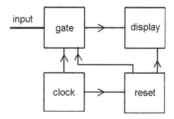

Figure 11.5:
Simplified block diagram for a frequency meter

tackled by using a switch for the master frequencies that allows harmonics for the crystal oscillator to be used. The problem of difficult multiples can be solved by counting both the master clock pulses and the input pulses, and operating the gate only when an input pulse and a clock pulse coincide. The frequency can then be found using the ratio of the number of clock pulses to the number of input pulses.

Suppose, for example, that the gate opened for 7 input pulses and in this time passed 24 clock pulses at 10 MHz. The unknown frequency is then $10 \times 7/24$ MHz, which is 2.9166 MHz. Frequency meters can be as precise as their master clock, so that the crystal control of the master oscillator determines the precision of measurements.

Counter/Timer

A counter/timer is used for the dual roles of counting pulses and measuring the time between pulses, and Figure 11.6 shows the block diagram, omitting reset and synchronizing arrangements.

The counter portion is a binary counter of as many stages as will be required for the maximum count value. The input to the counter is switched, and in the counting position, the input pulses operate the counter directly. When the switch is changed over to the timing position, the input pulses are used to gate clock pulses, and it is the clock pulses that are counted. In a practical circuit, the display would be changed over by the same switch so as to read time rather than count.

The timer action is obviously applied to digital watches and clocks. For both watch and clock applications the frequency of vibration of the quartz crystal will be around 32 kHz and this is counted down to suit the display, driving either a liquid crystal display (LCD) unit or a miniature motor to give 1 s impulses to a gear-train so as to make use of the traditional second, minute, and hour hand display. For a watch, the size of the electronics unit is all

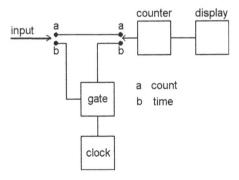

Figure 11.6:
Block diagram for a counter/timer instrument

important, particularly for ladies' watches, but larger watches and clocks have more space to play with. Watches are also limited for battery power, and the miniature button cells have a life that depends on the watch design, some providing five years or more.

A recent trend in both watches and clocks is synchronization from master signals such as the MSF Rugby time signals (now sent out from a transmitter in Cumbria). The US equivalent is station WWVB. The master clock at Rugby (similar to others around the world) uses the vibration of the caesium-133 atom as a time standard. The natural frequency is 9.192631770 GHz, and its precision is ± 1 second per million years. The signal is transmitted at a more modest frequency of 60 kHz and it carries coded information on month, day of month, day of week, hour, minute, and second, with additional codes that control the changes to and from summer time.

■ Note

MSF is not a technical abbreviation; it was the old call-sign for the Rugby radio transmitter which was built in 1923, using an antenna of about 43 km of copper cable (and 190 km of copper wire used as an earth connection).

Watches such as the Casio Waveceptor conserve battery power by switching on the MSF detector at one or more times in the night for long enough to download the information. Battery power is further conserved by moving the minute hand every 20 s rather than continually, so that battery life of five years or more can be obtained. Larger units in clocks can use AA cells and keep a continuous reception of MSF, along with a continuously moving second hand with gearing to provide minutes and hours.

The most useful feature of these units is that they can automatically respond to the hour advance in spring and the reverse action in autumn. The watches have the disadvantage of the older digital watches that they have to be controlled by buttons, and advancing or retarding the display on a flight can be a fiddly business.

Summary

Frequency meters and counter/timer circuits are obvious applications of counter circuits to measuring devices, and they have completely replaced older methods. The most familiar applications of timers are in quartz watches and clocks.

Digital Voltmeter Basics

At one time, all voltmeters were of the analog type. The voltage that was to be measured was used to pass current through a resistor of precise and known values, and the current passing was measured by a meter. The meter scale was calibrated in terms of volts, so that the reading was direct; it did not require any calculations. The problem with this type of meter is that it takes a current from the circuit, so that it alters, even if only slightly, the voltage that it is measuring. Another point is that the instrument is fragile, depending on a delicate meter movement. Modern digital meters take a negligible current and present the results as a number rather than as a scale reading. They are also more robust.

The most common IC of this type is the digital DC voltmeter IC, and a brief account of its action provides some idea of the operating principles of many instrumentation ICs.

Referring to the block diagram of Figure 11.7, the voltmeter contains a precision oscillator that provides a master pulse frequency. The pulses from this oscillator are controlled by a gate circuit, and can be connected to a counter. At the same time, the pulses are passed to an integrator circuit that will provide a steadily rising voltage from the pulses.

When this rising voltage matches the input voltage exactly, the gate circuit is closed, and the count number on the display then represents the voltage level. For example, if the clock frequency were 1 kHz, then 1000 pulses could be used to represent 1 V and the resolution of the meter would be 1 part in 1000, though it would take 1 s to read 1 V. After a short interval, usually around 0.25 s (determined by using another clock or a divider from the main clock) the integrator and the counter are reset and the measurement is repeated. Complete meter modules can be bought in IC form, using fast clock rates and with the repetition action built in.

Summary

The digital voltmeter is a less obvious use of counters in measurement, and its action depends on the use of a very precise integrator in IC form to produce a sawtooth voltage waveform from a set of counted pulses. Digital voltmeters have almost completely replaced the older analog type (which now turn up at antiques fairs).

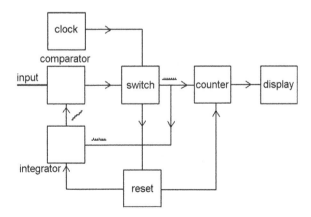

Figure 11.7:
Block diagram for a digital voltmeter. The block is for the main measuring unit, neglecting resistors that are used to extend the range for larger voltages

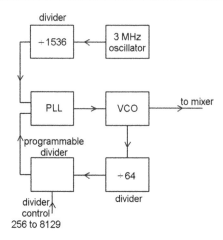

Figure 11.8:
A frequency synthesizer form of oscillator as used for television tuners

A very different application of digital methods is illustrated in Figure 11.8, which shows a **frequency synthesizer** such as was used on analog television receivers and in a simpler version for frequency-modulated (FM) receivers as a local oscillator for the superhet action. This, strangely enough, depends for its action on an **analog** circuit called a **phase-locked loop** (PLL), which compares the phases of two input signals of about the same frequency and whose output is a voltage whose size depends on the amount of phase difference between the input signals. If the input frequencies are different the output of the PLL is not a steady voltage, but if the frequencies are identical but with a phase difference, then the PLL output is a steady voltage.

The reference oscillator uses a crystal control and is precise enough to maintain a frequency as constant as any of the transmitted carriers. Its frequency is typically 3 MHz, which allows the use of comparatively low-cost components. This 3 MHz frequency is divided by a factor of 1536 to give 1953.125 Hz, and this frequency is used as one input to the PLL. Another oscillator, labeled VCO (voltage-controlled oscillator), is used as the local oscillator for the television receiver, and will be working at an ultrahigh frequency (UHF), such as 670.75 MHz.

The point of using the VCO is that its operating frequency can be altered by applying a steady input voltage, replacing any form of mechanical tuning. This oscillator frequency is divided by a fixed factor of 64 to give 10.480468 MHz, and this frequency is then used as the input of a programmable divider, whose division ratio can be anything from 256 to 8191. The numbers that control these division ratios are obtained from a memory within the tuner, and this is programmed with control number for each possible channel when the receiver is first set up. This system allows tuning to be much less affected by low signal strength, so that a radio with a digital tuner can be used where an older type would not lock on to a signal.

For example, if the frequency of 10.480468 MHz is divided by 5366, the result is 1953.125 Hz, which is identical to the frequency from the crystal-controlled oscillator. If the VCO frequency changes slightly, the frequencies into the PLL will no longer be equal, and a voltage will be generated from the PLL and used to correct the VCO. If you want to change channels, you press a switch that alters the divider ratio, causing a change in the PLL output that will then alter the frequency of the VCO until the new channel is perfectly tuned. This is all a long way from the variable-capacitor tuning of the early radios.

Summary

Frequency synthesizers are another application of counter circuits, and this type of action is now extensively used in radio and television receivers to produce the correct oscillator frequency for a superhet receiver circuit. This type of tuning is better able to use remote control and push-button selection actions, and has completely replaced the old-style rotating-knob tuning for television receivers, and FM or digital radio receivers.

Registers

Definition

The use of flip-flops in binary counters is just one application of these versatile building-blocks. A register, usually found in integrated form, is another method of using flip-flops for storage of binary digits and for logic operations that are called **shifting** and **rotating**.

A register is created from a set of flip-flops connected together, and there are four basic methods which are distinguished by the initials of their names. The simplest type is called parallel in, parallel out (PIPO), and a register of this type has no connections between the flip-flops (Figure 11.9).

Each flip-flop of this type of register has an input and an output, and the principle is that the inputs will determine the state of the outputs, and these outputs will be maintained until another set of input signals is used to change them. This type of register is used for storage, because each flip-flop is storing a bit (0 or 1) and will hold this bit until new input signals are used or until the power is switched off. This system was at one time used for computer memory, but it requires too much power (because each flip-flop draws current whether it is storing 1 or 0) for modern systems. The system, also called static random access memory (SRAM), is still used for small, fast memories.

The use of a register for very short-term storage is called **latching**. For example, using a latch register allows a set of digits that may exist for only one clock cycle (the time between two clock pulses) to be viewed on a display for as long as is needed. This facility is particularly useful for measuring instruments in which the signals may change rapidly but the display must

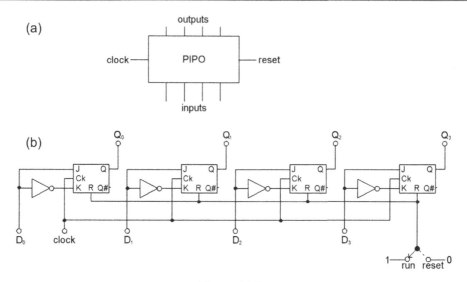

Figure 11.9:
The PIPO register, showing (a) the symbol and (b) a circuit using flip-flops

remain static for long enough to be read. A timer, for example, would be of little use if the time, measured in units of milliseconds, was displayed directly as the counter output, because the display on the least significant digits would be constantly changing and would not be readable until the timer stopped.

It is better to use a register as a latch between the counter and the display so that the time can be displayed without stopping the counter. This also allows intermediate times to be displayed. We can, for example, display the time for one lap of a race, while allowing the counter to continue timing the rest of the race. A latching action is also an essential part of a modern digital voltmeter (using designs more modern than the illustration in Figure 11.7), providing a steady reading (which is updated once per second) even when the voltage level is fluctuating slightly.

Summary

A register is a circuit that uses a set of flip-flops. The simplest type is the parallel in, parallel out register (PIPO) or latch, in which the flip-flops are independent (apart from a reset line), each with an input and an output. The action is used mainly for storing the bits of a digital signal, so that the register must use as many flip-flops as there are bits in the stored signal.

The opposite type of register is titled serial in, serial out (SISO). Each flip-flop has its output connected to the input of the next flip-flop in line (Figure 11.10), so that the whole register has just one input and one output. This is used as a signal delay and as a counter. On each clock pulse input, the bits in the flip-flops are shifted, meaning that they cause the next flip-flop in line to change. For a set of four flip-flops this will cause the waveforms to appear as shown in

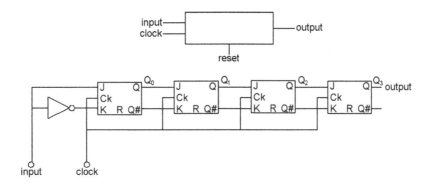

Figure 11.10:
The SISO register with symbol and a circuit using flip-flops

Figure 11.11. If there are no connections to the intermediate flip-flop outputs, this register acts as a divide-by-four or count to four circuit.

The parallel in, serial out (PISO) register is used mainly to convert parallel signals on a set of lines into a serial signal on one line. A set of bits on the input lines will affect each flip-flop in the register. By pulsing the clock line once for each flip-flop, each bit is fed out in turn from the serial output. The opposite type is the serial in, parallel out (SIPO) register, in which bits are fed in one at a time on each clock pulse and appear on the parallel output after a number of clock pulses that is equal to the number of flip-flops in the register.

You never need to construct registers from individual flip-flops, because all forms of registers are obtainable in completely integrated form, often as units that can be connected so as to obtain whatever action you want. Integrated registers often feature shift direction controls so that by altering the voltage on a pin, you can switch between left-shift and right-shift. When the input of a serial register is connected to the final output, the action is **rotation**; the bits are cycled round one stage for each clock pulse.

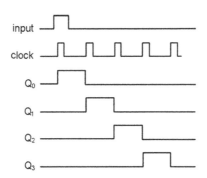

Figure 11.11:
The result of using a square pulse input for the first of a set of clock pulses. The pulse is cycled through the register by one stage on each clock pulse

Summary

The registers with serial input or output are used extensively for converting between serial and parallel data signals, and are an alternative to the use of multiplexers and demultiplexers (see Chapter 10). They are also used in multiplication and division actions.

Related Chips

Microprocessors and microcontrollers are the best known types of large-scale logic devices, but there is a large family of related devices that we need to be aware of, even if we seldom encounter some of them. These devices are combinations of gates, and the inevitable rapid development of construction methods has resulted in a large number of new names. This is more specialized territory, so we will look at some of the most important ones without going into detail.

SPLD, CPLD, and FPGA

The simple programmable logic device (SPLD) consists of a set of gates and a flip-flop. One SPLD can therefore replace a small logic circuit that can have several inputs and one output. These chips can be sold under various names, such as programmable logic device (PLD), programmable logic array (PLA), field programmable logic array (FPLA), generic array logic (GAL), and programmable array logic (PAL).

The CPLD is a much more complex variety of SPLD, replacing thousands of logic gates in one package. Equipment can be used to set up the connections in the CPLD to carry out the required actions.

The field-programmable gate array (FPGA) is, as the name suggests, a set of gates on one chip for which connections can be made by the user (rather than by the manufacturer), usually after the chip has been soldered to a board. FPGAs are more likely to be found in complex circuits such as microprocessor support systems.

ASIC and ASSP

The name ASIC means application-specific integrated circuit, meaning that the internal circuits are designed and connected up to carry out one specific task for a single company in a specific application. ASICs offer high performance and low power consumption compared to a circuit using off-the-shelf components, but their internal design is very complex so that they are both time-consuming and expensive to create. Many of today's digital devices, ranging from mobile phones to games consoles, depend heavily on ASICs.

A smaller scale version of ASIC is the application-specific standard product (ASSP). This is closer to a general-purpose IC, but the technology is closely based on that for ASICs.

SoC

SoC means system on a chip, and it is a single chip of the ASIC or ASSP type that contains all the components and connections of a complex system (such as a computer with microprocessor or microcontroller, memory, peripherals, logic, etc.) into a single integrated circuit. This need not be an all-digital system, because the SoC may contain analog processing and even radio frequencies; an example is the SoC used for digital audio broadcasting (DAB) radio receivers.

Memory Chips

Microcontrollers often include memory actions, but microprocessors are more likely to need separate memory chips, because for computing purposes memory sizes in the region of 1 GB are now almost a minimum requirement. This type of memory is called random access memory (RAM) and we will deal with it in Chapter 13.

RAM is memory that retains data only while power is applied, but most microprocessor applications need some permanent memory that will hold data without the need for a power supply. This general type of memory is called read-only memory (ROM), implying that its contents are fixed and cannot be altered. The simplest type consists of a CPLD with a fixed set of connections created in the manufacturing process.

Sometimes, however, we need other types of memory. For some applications, the content of fixed memory needs to be determined by the customer or by an engineer testing a new device. Two devices answer this requirement: erasable programmable read-only memory (EPROM) and EEPROM. The EPROM is a memory chip with a 'window' that is transparent to ultraviolet (UV) light. Shining UV light on to this window will erase connections that have been programmed into it by applying signals to the pin connections. The chip can then be reprogrammed with a different set of internal connections, hence a different set of stored bytes of data.

The EEPROM is a later development that allows the erasure to be carried out by electrical (hence the extra E) signals at a voltage higher than is used for reading. Once again, this type of ROM is very useful when circuits are being developed, because it allows ideas to be tried out without the need to manufacture an expensive one-off memory.

Flash memory is a much later development of EEPROM that does not require higher voltages; it simply uses the signal at one contact to determine whether it is reading or writing. This type of memory is now almost universally used in familiar devices such as the camera secure data (SD) card and the universal serial bus (USB) flash memory stick.

Error Control and Correction

Any system that uses digital signals is much less likely to be upset by noise than the equivalent analog signal. Consider, for example, the square-wave signal in Figure 11.12, which has been degraded by noise, and which can be 'cleaned up' simply by slicing the received signal at the two extremes of voltage levels. This type of cleaning up is impossible when an analog signal is being used, because there is no simple way of distinguishing the noise from the analog signal itself.

Figure 11.13 is a graph illustrating how these two types of signals behave when the signal-to-noise (S/N) ratio decreases by the same amount. The analog system is affected progressively, but the digital system is almost unaffected by the noise until a point is reached, when the system suddenly crashes as the bit error rate (BER) rises. This does not mean that digital methods cannot be used when the noise level is high, because there are a number of ways in

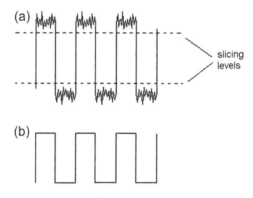

Figure 11.12:
Noise immunity of a digital signal. The noise affects the tips of the waveform, which can be squared by clipping (slicing) at suitable levels

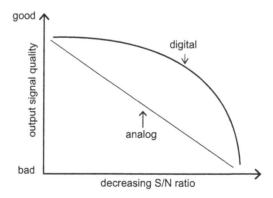

Figure 11.13:
Comparison of noise immunity for comparable analog and digital circuits

which the information can be accurately recovered even from signals with a high BER. These methods are always used in digital recording systems [such as compact discs (CDs)] and for transmission systems (digital telephone links) and are the main reason for the widespread adoption of digital methods for signaling. A typical bit error rate for noisy conditions is 1 error in 1000 (a BER of 10^{-3}), and for better conditions, about 1 in 10,000 (BER = 10^{-4}).

■ Note

The effects of high BERs are very noticeable on digital television or radio broadcasting. Whereas a noisy analog signal provides a speckled picture of noisy sound, a digital signal with a high error rate can result in a still picture or a blank screen, and sound that is intermittent. The BER for television is usually high enough to avoid such problems, but DAB radio suffers considerably more because of poor transmitter coverage and inadequate antennae.

■

Summary

All forms of electronic signals are degenerated by noise. For analog signals, this is expressed as the signal-to-noise (S/N) ratio, and this figure should be high, typically 40 dB or more. For digital signals, the effect is expressed as bit error rate (BER), and this should be low, of the order of 1 in 1000 (10^{-3}) or 1 in 10,000 (10^{-4}) if data correction methods are to work well. The important difference between analog and digital methods is that whereas analog signals gradually deteriorate, digital signals are almost unaffected until the noise reaches a critical level, when the digital signal becomes totally garbled.

The simplest type of system that can be used for correcting signals sent from a transmitter to a receiver is known as automatic request for repeat (ARQ). A standard code (American Standard Code for Information Interchange or ASCII) has for many years been used for teleprinters and computers. This uses two special code numbers that are referred to as ACK and NAK. If a distant receiver detects that a code pattern has no errors, it transmits back the ACK (Acknowledge) code. If errors have been detected, the NAK (Negative acknowledge) code is returned, and this is used to make the transmitter repeat the transmission of the last block of signal code.

A more advanced system is known as forward error control (FEC), and we have looked already at one version of this in Chapter 9, the use of parity bits. All FEC systems make use of the addition to the signal bits of extra redundant bits which, when suitably processed, are capable of identifying the errors and, in some cases, correcting the errors.

The main causes of bit errors in digital systems (and loss of signal in analog systems) are **white noise** and **impulsive noise**. White noise consists of a mixture of frequencies that is fairly evenly spread over the frequency range; no frequency is affected more than any other. This type of noise produces errors that are random. Impulsive noise, such as is produced by car-ignition systems, is at some definite pulse rate and it can create bursts of errors on each pulse. The errors that can occur are of three types, classed as detectable and correctable, detectable but not correctable, and undetectable (and therefore uncorrectable).

If an error can be detected but not corrected, it is possible that the error can be concealed in some way so that it causes less damage. For example, the system could ignore the error in a number and treat it as a zero level, or it could make use of the last known correct value again or it could calculate an average value based on the previous and next correct values. This last method is called **interpolation**.

Using a Check Sum

The check sum is an error control scheme that is often used for digital recording systems, particularly magnetic disc and tape. The data must be organized into long 'blocks', each of which can be addressed by a reference number. For each block of digital data (numbers) the digital sum of the numbers, the **check sum**, in a block is stored at the end of that block. When the data is recovered, the check sum can be recalculated from the block of replayed numbers and compared with the original to find whether any errors have occurred during reading or storage.

The simple check-sum technique can only identify when an error has occurred and cannot indicate whereabouts in the block the error is located. A modification of this method is called the **weighted check sum** method, and it relies on the behavior of prime numbers (numbers that have no factors other than 1), such as 1, 3, 5, 7, 11, 13, etc. In a weighted check-sum system, each data number in a block is multiplied by a different prime number, and the check sum is calculated from this set of numbers. If a different check sum is found on replay, the difference between the calculated and the stored check sum is equal to a prime number, the same prime number as was used to multiply the data number that has been corrupted, so locating the number. The mathematical basis of this system is beyond the scope of this book but it is fascinating to see how an apparently academic study such as that of prime numbers can be put to a very practical engineering application. In mathematics and science, there is no such thing as useless knowledge, just knowledge that we do not yet need or know that we may need.

Summary

The use of ACK and NAK codes along with parity is a simple system for correcting transmitted data, but nowadays the use of systems such as check sums is more usual. The more advanced systems deal with a block of data numbers at a time, and can detect an error in the block, and correct this error if the coding allows its position to be located.

Hamming and Other Codes

Hamming codes are a form of error-correcting codes that were invented by R.W. Hamming, born in 1915, the pioneer of error-control methods. The details of Hamming codes are much too mathematical for this book, but the principles are to add check-bits to each binary number so that the number is expanded; for example, a 4-bit number might have three check-bits added to make the total number size seven bits. The check-bits are interleaved with the number bits, for example by using positions 1, 2, and 4 for check-bits, and positions 3, 5, 6, and 7 for the bits of the data number.

When an error has occurred, the parity checking of the check-bits will produce a number (called a **syndrome**) whose value reveals the position of an error in the data. This error can then be corrected by inverting the bit at that position. A simple Hamming code will detect a single bit error, but by adding another overall parity bit, double errors can be detected. Hamming codes are particularly useful for dealing with random errors (caused by white noise).

The **cyclic redundancy check** (CRC) method is particularly effective for dealing with burst errors caused by impulsive noise, and is extensively used for magnetic recording of data. Once again, the mathematical details are beyond the scope of this book, but in general, a data block is made up from three words, the data word, a generator code word, and a parity check code word. The data word is divided by the generator word to give the parity check word, and the whole block is recorded. On replay, the generator code is separated, and the whole block number is divided by this code. If the result is zero, there has been no error. If there is a remainder after this division, the error can be located by using this remainder along with the parity code.

The basic Hamming and CRC systems have been improved and developed over many years to produce digital systems that are more effective in dealing with both random errors and bursts of errors. These systems include BCH codes for random error control, Golay codes for random and burst error control, and Reed–Solomon codes for random and very long burst errors with an economy of parity bits. The Reed–Solomon coding system is used for CD recording and replay.

Summary

The more advanced methods of error detection and correction, ranging from simple Hamming codes through CRC to Reed–Solomon, all make use of added bits in a block of data and mathematical methods that allow the position of an error to be found. These methods also allow for error correction.

CHAPTER 12

Compact Disc System and Beyond

Compact Discs

The compact disc (CD) system is a rare example of international cooperation leading to a standard that has been adopted worldwide, something that was notably lacking in the first generation of video-cassette recorders. When video-cassette recorders first appeared they were performing an action that was not possible before, but audio compact discs would be in competition with existing methods of sound reproduction, and it would have been ridiculous to offer several competing standards. The CD was the first consumer product to feature digital coding as a replacement for analog methods, and though the system still looks and sounds impressive it is now being overtaken by more recent developments. Nevertheless, the principles of CD recording are likely to be with us for a considerable time.

■ Note

In addition, because the CD was the first consumer product to use digital methods, the principles used in the CD system have been applied to later products such as digital versatile discs (DVDs), the short-lived Minidisc, MP3, cinema sound, and in the digital broadcasting of sound and pictures.

■

In 1978, a convention dealing with digital disc recording was organized by some 35 major (mainly Japanese) manufacturers who recommended that development work on digital discs should be channeled into 12 directions, one of which was the type proposed by Philips. The outstanding points of the Philips proposal were that the disc should use constant linear velocity recording, eight-to-fourteen modulation, and a new system of error correction called cross-interleave Reed–Solomon code (CIRC).

Definition

Constant velocity recording means that no matter whether the inner or the outer tracks are being scanned, the rate of digital information should be the same, and Philips proposed to do this by varying the speed of the disc depending on the distance from the center of the track that was being read, so that the rate of reading was constant.

Electronics Simplified. DOI: 10.1016/B978-0-08-097063-9.10012-3

■ **Note**

You can see the effect of this for yourself if you use a CD player in which the disc is visible. Play a track from the start of a CD (the inside tracks) and then switch to a track at the end (outside). The reduction in rotation speed when you make the change is very noticeable, and this use of constant digital rate makes for much simpler processing of the data.

■

By 1980, Sony and Philips had decided to pool their respective expertise, using the disc modulation methods that had been developed by Philips along with signal-processing systems developed by Sony. The use of a new error-correcting system along with higher packing density made this system so superior to all other proposals that companies flocked to take out licenses for the system. All opposing schemes died off, and the CD system emerged as a rare example of what can be achieved by cooperation in what is normally an intensely competitive market.

Summary

The audio CD system is a global standard, obtained by pooling the expertise of Philips and Sony. From the start, the system was designed to produce much higher performance than was obtainable using analog methods, and provision was made for other uses, allowing CD to be used in later developments such as CD-ROM for computers, then in developments such as DVD and Blu-ray.

The Optical System

The CD system makes use of optical recording, using a beam of light from a miniature semiconductor laser. Such a beam is of low power, a matter of milliwatts, but the focus of the beam can be to a very small point, about 0.6 μm in diameter; for comparison, a human hair is around 50 μm in diameter. The beam can be used to form pits in a flat surface, using a depth that is also very small, of the order of 0.1 μm. If no beam strikes the disc, then no pit is formed, so that we have here a system that can digitally code pulses into the form of pit or no-pit. These pits on the master disc are converted to pits of the same scale on the copies. The pits or their opposite form of dimples are of such a small size that the tracks of the CD can be much closer: about 60 CD tracks take up the same width as one vinyl LP track.

■ **Note**

A pit is as readable as a dimple: either will cause the light beam to scatter instead of being reflected back, so that it does not matter which way the surface is changed by the presence of a '1' bit. We will see later that a perfectly flat disc can be used if the bits are recorded as color or phase changes.

■

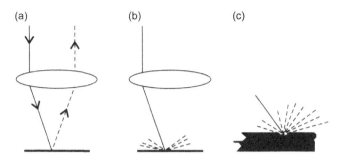

Figure 12.1:
The compact disc. (a) Light reaching a flat portion is reflected back; (b) light striking a pit is scattered so that there is no strong reflected beam; (c) a close-up

Reading a set of pits/dimples on a disc also makes use of a semiconductor laser, but of much lower power, since it need not vaporize the material. The reading beam will be reflected from the disc where no pit exists, but scattered where there is a pit (Figure 12.1). By using an optical system that allows the light to travel in both directions to and from the disc surface (Figure 12.2), it is possible to focus a reflected beam on to a photodiode, and pick up a signal when the beam is reflected from the disc, with no signal when the beam falls on to a pit/dimple. The output from this diode is the digital signal that will be amplified and then processed eventually into an audio signal. Only light from a laser source can fulfill the requirements of being perfectly **monochromatic** (one single frequency) and **coherent** (no breaks in the wave-train) so as to permit focusing to such a fine spot.

The CD player uses a beam that is focused at quite a large angle, and with a transparent coating over the disc surface that also focuses the beam as well as protecting the recorded pits. Though the diameter of the beam at the pit is around 0.5 µm, the beam diameter at the surface of the disc, the transparent coating, is about 1 mm. This means that dust particles and hairs on the surface of the disc have very little effect on the beam, which passes on each side of them, unless your dust particles are a millimeter across! This, illustrated in Figure 12.3, is just one of the several ways in which the CD system establishes its very considerable immunity to dust and scratching on the disc surface, the other being the advanced bit-error detection and correction system.

■ Note

Given, then, that the basic record and replay system consists of using a finely focused beam of laser light, how are the pits or dimples arranged on the disc and how does the very small reading spot keep track of the pits? To start with, the pits are arranged in a spiral track, like the groove of a conventional record. This track, however, starts at the **inside** of the disc and spirals its way to the outside, with a distance between adjacent tracks of only 1.6 µm. Since there is no mechanical contact of any kind, the tracking must be carried out by a **servomotor** system that accurately guides the laser and lens assembly.

■

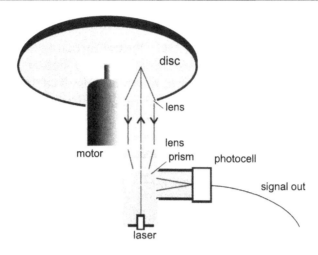

Figure 12.2:
The optical arrangement of a compact disc (CD) player. The motor spins the disc, and the laser light is focused to a spot which is reflected where there is no pit or dimple on the disc. The reflected light is diverted by the prism and falls on the photocell, causing an output

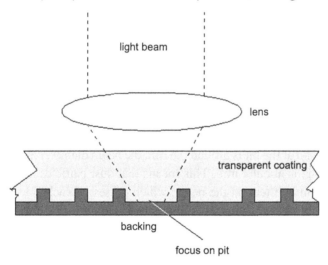

Figure 12.3:
Illustrating that the beam has a larger radius as it passes through the transparent coating. This avoids errors due to surface scratches. The ratio of beam diameter at the transparent surface to beam diameter at the reflective surface is enormous: about 2000 times

Definition

A servomotor is a small electric motor that is used in a feedback loop. Electronic sensors detect that there is an error in position, and operate the servomotor until the error is zero, at which point there is no output to the servomotor.

The principle is that the returned light will be displaced if there is mistracking, and the detector unit contains two other photodiodes to detect signals that arise from mistracking in either direction. These signals are obtained by splitting the main laser beam to form two side-beams, neither of which should ever be deflected by the pits on the disc when the tracking is correct. These side beams are directed to the photodiodes. The signals from these diodes are used to control the servomotors in a feedback loop system, so that the corrections are being made continually while the disc is being played.

In addition, the innermost track carries the position of the various music tracks in the usual digitally coded form. Moving the scanner system radially across the tracks will result in signals from the photodiode unit, and with these signals directed to a counter unit, the number of tracks from the inner starting track can be counted. This therefore allows the scanner unit to be placed on any one of the possible 41,250 tracks. This is a precision that could hardly have been dreamt of in the days of vinyl recordings.

Positioning on to a track is a comparatively slow operation because part of each track will be read on the way before the count number is increased. In addition, the motor that spins the disc is also servo controlled so that the speed of reading the pit/dimples is constant. Since the inner track has a diameter of 50 mm and the maximum allowable outer track can have a diameter of 116 mm, the number of pits per track will also vary in this same proportion, about 2.32. The disc rotation at the outer edge is therefore slower by this same factor when compared to the rotational speed at the inner track. The rotational speeds range from about 200 r.p.m. on the inner track to about 500 r.p.m. on the outer track. The actual values depend on the servo control settings and do not have to be absolute in the way that the old $33\frac{1}{3}$ r.p.m. speed for vinyl discs had to be absolute. The only criterion of reading speed on a CD is that the pits/dimples are being read at a correct rate, and that rate is determined by a master clock frequency. This corresponds to a reading speed of about 1.2–1.4 m of track per second.

The next problem is of recording and replaying two channels for a stereo system, because the type of reading and writing system that has been described does not exactly lend itself well to twin-channel use with two tracks being read by two independent scanners. This mechanical and optical impossibility is avoided by recording the two-channel information on a single track, recording samples from the channels alternately.

This is made possible by the storage of data while they are being converted into blocks called **frames** for recording, because if you have one memory unit containing left-channel data and one containing right-channel data, they can be read alternately into a third memory ready for making up a complete frame. There is no need to have the samples taken in a phased way to preserve a time interval between the samples on different channels as long as the signals are stored long enough to allow matched left and right signals to be output. As the computer jargon puts it, the alternation does not need to take place in **real time**. This is another advantage that digital methods have over analog methods.

Summary

The optical system that is used for recording or reading CDs is based on low-power semiconductor lasers whose beam can be focused to a very small point. On recording, each logic 1 is reproduced as a pit or dimple on the finished CD. For reading, each pit scatters the laser beam so that it is not reflected back to a detector. The guidance system uses two other detectors which produce a constant output while the main beam is correctly aimed. This is achieved using a servomotor system with the error signals used to correct the position, and the same system is used to move from one part of the continuous track to another. Stereo recording is achieved by recording left and right data alternately, using storage so that the signals can be reassembled correctly.

Digital Signal Processing

We now have to deal with the most difficult parts of the whole CD system. These are the modulation method, the error-detection and correction, and the way that the signals are assembled into frames for recording and replay. The exact details of each of these processes are released only to licensees who buy into the system. This is done on a flat-rate basis: the money is shared out between Philips and Sony, and the licensee receives a manual called the 'Red Book' which precisely defines the standards and the methods of maintaining them. Part of the agreement, as you might expect, is that the confidentiality of the information is maintained, so that what follows is an outline only of methods. This is all that is needed, however, unless you intend to design and manufacture CD equipment, because for servicing work you do not need to know the circuitry of an integrated circuit (IC) or the content of a memory in order to check that it is working correctly. It is important to know about these methods, because they spring up again when we look at digital television and radio. In addition to the Red Book, which defines the use of CD for audio, there are various other specifications (Orange Book, White Book, etc.) that define how the type of CD can be used for video and for computer data.

■ Note

Digital signal processing or digital signal processor (DSP) means in general the use of numbers (digits) to represent electrical signals and to carry out actions on such signals. The processing that is applied to create a CD is just one example of DSP. Many types of signals inevitably start out as analog; music is just one example. In the past we processed these signals by analog methods, but this imposes limitations (such as distortion) that we looked at earlier in this book. For many of the applications where analog signals are generated, we can convert these signals into digital versions and use a microprocessor to work on the digital signals. A dedicated type of microprocessor for this type of action is the DSP, and we can think of it as an enhanced microprocessor intended to carry out digital signal processing more efficiently.

Definition

A long-burst error is the type of error in which a considerable amount of the signal has been corrupted. This is the type of error that can be caused by scratches on the disc; it is long in comparison to the size of the scanning beam.

The error-detection and correction system is called CIRC. This code is particularly suitable for the correction of long-burst errors, and the code that is used for CDs can handle errors of up to 4000 bits, corresponding to a 2.5 mm fault on the disc lying along the track length. A coding that required too many redundant bits to be transmitted would cramp the expansion possibilities of the system, so the CIRC uses only one additional bit added to each three pieces of data; this is described as an efficiency of 75%. Small errors can be detected and remedied, and large errors are dealt with by synthesizing the probable waveform on the basis of what is available in the region of the error.

■ Note

The whole system depends heavily on a block or **frame** structure which, though not of the same form as a television signal frame, allows for the signal to carry its own synchronization pattern. This makes the recording **self-clocking**, so that all the information needed to read the signal correctly is contained on the disc; there is no need, for example, for each player to maintain a crystal-controlled clock pulse generator working at a fixed frequency. In addition, the use of a frame structure allows the signal to be recorded and replayed by a rotating head type of tape recorder, so that the digital signals from a master recorder can be transferred to a CD without the need to convert to analog and back.

Each frame of data contains a synchronization pattern of 24 bits, along with 12 units of data (using 16-bit words of data), four error-correcting words (each 16 bits) and 8 bits of control and display code. Of the 12 data words, six will be left-channel and six right-channel, and the use of a set of 12 in the frame allows expansion to four channels (three words each) if needed later. Though the actual data words are of 16 bits each, these are split into 8-bit units for the purposes of assembly into a frame. The content of the frames, before modulation and excluding synchronization, is therefore as shown in Table 12.1.

Summary

The digital signals are assembled into a **frame,** along with synchronization bits, error-correcting words, and control codes, amounting to 264 bits per frame. This is the basic digital unit, and it is then modulated into its final format.

Table 12.1: How the signals are organized into a frame of data: (a) as bytes, (b) after 8 to 14 modulation (EFM).

(a) As bytes			
Data	12 × 16 bits	= 24 × 8 bits	
Error correction	4 × 16 bits	= 8 × 8 bits	Total 33 × 8 bits
Control/display		1 × 8 bits	
		264 bits per frame	
(b) After EFM			
Total data of 33 bytes	33 × 14 bits	= 462 bits	
Synchronization		= 24 bits	
Redundant bits	3 × 24 bits	= 102 bits	
Total		588 bits per frame	

EFM

Definition

EFM means eight-to-fourteen modulation, in which each set of eight bits is converted to a set of fourteen bits for the purposes of recording. There is no mathematical way of converting, and the action is carried out by using a **lookup table**, a form of memory chip, in which each 8-bit number has a 14-bit counterpart. The 14-bit numbers are carefully chosen so that they avoid long runs of the same bit, or rapid alternations between bits.

The word **modulation** is used here to mean the method that is used to code the 1s and 0s of a digital number into 1s and 0s for recording purposes. The principle is to use a modulation system that will prevent long runs of either 1s or 0s from appearing in the final set of bits that will burn in the pits on the disc, or be read from a recorded disc, and calls for each digital number to be encoded in a way that is quite unlike binary code. The system that has been chosen is EFM.

The purpose of EFM is to ensure that each set of 8 bits can be written and read with minimum possibility of error. The code is arranged, for example, so that changes of signal never take place closer than 3 recorded bits apart. This cannot be ensured if the unchanged 8- or 16-bit signals are used, and the purpose is to minimize errors caused by the size of the beam, which might overlap two pits and read them as one. The 3-bit minimum greatly eases the requirements for perfect tracking and focus.

At the same time, the code allows no more than 11 bits between changes, so avoiding the problems of having long runs of 1s or 0s. In addition, three redundant bits are added to each 14, and these can be made 0 or 1 so as to break up long strings of 0s or 1s that might exist when two blocks of 14 bits were placed together. The conversion can be carried out by a small piece of fixed memory (ROM) in which using an 8-bit number as an address will give data output on 14 lines. The receiver will use this in reverse, feeding 14 lines of address into a ROM to get 8 bits of number out.

The use of EFM makes the frame considerably larger than you would expect from its content of 33 8-bit units. For each of these units we now have to substitute 14 bits of EFM signal, so that we have 33×14 bits of signal. There will be 3 additional bits for each 14-bit set, and we also have to add the 24 bits of synchronization and another 3 redundant bits to break up any pattern of excessive 1s or 0s here. The result is that a complete frame as recorded needs 588 bits, as detailed in Table 12.1. All of this, remember, is the coded version of 12 words, six per channel, corresponding to six samples taken at 44.1 kHz, and so representing about 136 μs of signal in each channel.

Summary

Normal binary signals, as assembled into a frame, are liable to contain rapid alternations (such as 1010101..) and also long sequences of the same bit type, and neither of these is acceptable for recording purposes. By expanding each 8-bit unit to 14 bits, the code can be changed so that it avoids rapid alternations and long sequences. This EFM (eight-to-fourteen modulation) code then has other bits added for separation and synchronization to make a complete frame occupy 588 bits for 12 words of data.

Error Correction

Definition

The errors in a digital recording and replay system can be of two main types, random errors, and burst errors. Random errors, as the name suggests, are errors in a few bits scattered around the disc at random and, because they are random, they can be dealt with by relatively simple methods. A burst error is quite a different beast, and is an error that involves a large number of consecutive bits.

Randomness also implies that in a frame of 588 bits, a bit that was in error might not be a data bit, and the error could in any case be corrected reasonably easily. Even if it were not, the use of EFM means that an error in one bit does not have a serious effect on the data. The EFM system is by itself a considerable safeguard against error, but the CD system needs more than this to allow, as we have seen earlier, for scratches on the disc surface that cause long error sequences. The main error correction is therefore done by the CIRC coding and decoding, and one strand of this system is the principle of **interleaving**. Before we try to unravel what goes on in the CIRC system, then, we need to look at interleaving and why it is carried out.

A burst or block error could be caused by a bad scratch in a disc or a major dropout on tape, and its correction is very much more difficult than the correction of a random error. Now if the bits that make up a set were not actually placed in sequence on a disc, then block errors would have much less effect. If, for example, a set of 24 data units (bytes) of 8 bits each that belonged together were recorded so as to be on eight different frames, then all but very large block errors would have much the same effect as a random error, affecting only one or two of the byte units. This type of shifting is called **interleaving**, and it is the most effective way of dealing with large block errors. The error-detection and correction stages are placed between the channel alternation stage and the step at which control and display signals are added prior to EFM encoding.

The CIRC method uses the Reed—Solomon system of parity coding along with interleaving to make the recorded code of a rather different form and different sequence from the original code. Two Reed—Solomon coders are used, each of which adds four parity 8-bit units to the code for a number of 8-bit units. The parity system that is used is a very complicated one, unlike simple single-bit parity, and it allows an error to be located and signaled. The CD system uses two different Reed—Solomon stages, one dealing with 24 bytes (8-bit units) and the other dealing with 28 bytes (the data bytes plus parity bytes from the first one), so that one frame is processed at a time.

In addition, by placing time delays in the form of serial registers between the coders, the interleaving of bytes from one frame to another can be achieved. The Reed—Solomon coding leaves the signal consisting of blocks that consist of correction code(1), data(1), data(2) and correction code(2), and these four parts are interleaved. For example, a recorded 32-bit signal may consist of the first correction code from one block, the first data byte of the adjacent block, the second data byte of the fourth block, and the second correction code from the eighth data block. These are assembled together, and a cyclic redundancy check number can be added.

At the decoder, the whole sequence is performed in reverse. This time, however, there may be errors present. We can detect early on whether the recorded 'scrambled' blocks contain errors and these can be corrected as far as possible. When the correct parts of a block (error code, data, data, error code) have been put together, then each word can be checked for errors, and these corrected as far as possible. Finally, the data is stripped of all added codes, and will either be error free or have caused the activation of some method (such as interpolation) of correcting or concealing gross errors.

Summary

Further error protection is achieved by interlacing frames, so that an error is less likely to affect a set of consecutive frames. The Reed—Solomon system provides a very considerable amount of detection and correction ability, and if all else fails, the signal will be interpolated, meaning that the missing digits will be replaced by values between the surrounding values. For example, in the sequence 1, 2, 3, 6, 7, 8, interpolation would fill in values of 4 and 5.

Control Bytes

For each block of data one control byte (8 bits) is added. This allows eight channels of additional information to be added, known as sub-codes P to W. This corresponds to one bit for each channel in each byte. To date, only the channels P and Q have been used. The P channel carries a selector bit which is 0 during music and lead-in, but is set to 1 at the start of a piece, allowing a simple but slow form of selection to be used. This bit is also used to indicate the end of the disc, when it is switched between 0 and 1 at a rate of 2 Hz in the lead-out track.

The Q channel contains more information, including track number and timing. A channel word consists of a total of 98 bits, and since there is one bit in each control byte for a particular channel the complete channel word is read in each 98 blocks. This word includes codes that can distinguish between four different uses of the audio signals.

The allocation of these control channels allows considerable flexibility in the development of the CD format, so that players in the future could make use of features that have to date not been thought of. Some of the additional channels are used on the CD-ROMs that are employed as removable storage for computers.

Production Methods

A commercial CD (as distinct from one produced by a computer) starts as a glass plate that is ground and polished to 'optical' flatness, meaning that the surface contains no deformities that can be detected by a light beam. If a beam of laser light is used to examine a plate like this, the effect of reflected light from the glass will be to add to or subtract from the incident light, forming an **interference pattern** of bright and dark rings. If these rings are perfectly circular then the glass plate is perfectly flat, so that this can be the basis of an inspection system that can be automated.

The glass plate is then coated with **photoresist**, a material that hardens on exposure to light (as in the old gum-bichromate photographic process). Various types of photoresist have been developed for the production of ICs, and these are capable of being printed with very much finer detail than is possible using the older type of photoresist which is used for printed circuit boards, and the thickness and uniformity of composition of the resist must both be very carefully controlled.

The image is then produced by treating the glass plate as a CD and writing the digital information on to the photoresist with a laser beam. Once the photoresist has been processed, the pattern of pits will develop and this comprises the glass master. The surface of this disc is then silvered so that the pits are protected, and a thicker layer of nickel is then plated over the surface. This layer can be peeled off the glass and is the first metal master. The metal master is used to make ten 'mother plates', each of which will be used to prepare the stamper plates.

Once the stamper plates have been prepared, mass production of the CDs can start. The familiar plastic discs are made by injection molding (the word 'stamper' is taken from the corresponding stage in the production of black vinyl discs) and the recorded surface is coated with aluminum, using vacuum vaporization. Following this, the aluminum is protected by a transparent plastic which forms part of the optical path for the reader and which can support the disc label. The system is illustrated in outline in Figure 12.4.

Summary

The CDs are produced by using photoresist material on a flat glass plate. This is then recorded using a laser that will harden the photoresist so that after processing a master can be produced. This is used to produce plates which will stamp out plastic copies.

The End Result

All of this encoding and decoding is possible mainly because we can work with stored signals, and such intensive manipulation is tolerable only because the signals are in the form of digital code. Certainly any efforts to correct analog signals by such elaborate methods would not be welcome

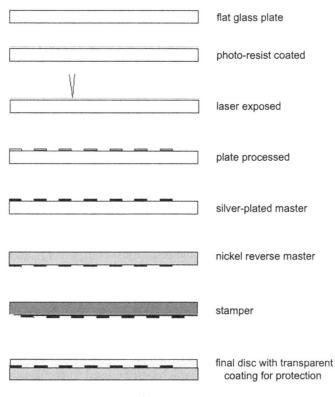

flat glass plate

photo-resist coated

laser exposed

plate processed

silver-plated master

nickel reverse master

stamper

final disc with transparent coating for protection

Figure 12.4:
Production of compact discs (CDs) from glass plate stage

and would hardly add to the fidelity of reproduction. Because we are dealing with signals that consist only of 1s and 0s, however, and which do not provide an analog signal until they are converted, the amount of work that is done on the signals is irrelevant. This is the hardest point to accept for anyone who has been brought up in the school of thought that the less that is done to an audio signal the better it is. The whole point about digital signals is that they can be manipulated as we please provided that the underlying number codes are not altered in some irreversible way.

The specifications for the error-correcting system are:

- maximum correctable burst error length is 4000 bits (2.5 mm length)
- maximum interpolable burst error length is 12,300 bits (7.7 mm length).

Other factors depend on the bit error rate, the ratio of the number of errors received to the number of bits in total. The system aims to cope with bit error rates (BERs) between 10^{-3} and 10^{-4}.

- For BER $= 10^{-3}$ the interpolation rate is 1000 samples per minute, with undetected errors less than 1 in 750 hours.
- For BER $= 10^{-4}$ the interpolation rate is 1 sample in 10 hours with negligible undetected errors.

Performance

The proof of the efficacy of the whole CD system lies in the audio performance. The bare facts are impressive enough, with a frequency range of 20 Hz to 20 kHz (within 0.3 dB) and more than 90 dB dynamic range, signal-to-noise ratio and channel separation figures, with total harmonic distortion (including noise) less than 0.005%. A specification of that order with analog recording equipment would be difficult, to say the least, and one of the problems of digital recording is to ensure that the analog equipment that is used in the signal processing is good enough to match up to the digital portion.

Added to the audio performance, however, we have the convenience of being able to treat the digital signals as we treat any other digital signals. The inclusion of control and display data means that the number of items on a recording can be displayed, and we can select the order in which they are played, repeating items if we wish. Even more impressive (and very useful for music teachers) is to ability to move from track to track, allowing a few notes to be repeated or skipped as required. The other consequence is that a CD can contain any data that can be put into digital form, so that it can carry sound, pictures, and text. This is the basis of multimedia work, and what we have seen to date only scratches the surface of what is possible.

Computer Optical Drives

The successful use of CDs for audio paved the way for using the same media for distributing digital data for other purposes such as computer programs and text or image data. Initially, optical discs were of the pressed type only, providing around 640 Mbytes of data storage, so that they had to be manufactured by the master-copy process that has been outlined for CDs. This formed an ideal way to sell computer programs. All of this changed when recordable CDs were devised.

The original CD standards that were drawn up by Philips and Sony were for audio, but the extended uses for other forms of digital data were anticipated, and by 1988 the concept of the writable CD was firmly established. The use of laser-formed pits/dimples had to be replaced by something that could be used with lasers of much lower writing power, and the scheme that eventually emerged uses layers of light-sensitive dyes. A writable CD (CD-R) starts off as a plain plastic disc that is produced from a mold, but the mold contains a spiral track with no pits or dimples. A layer of light-sensitive dye is deposited on the disc, and is then coated with a thin layer of metal and a thicker layer of transparent lacquer. The action of a writing laser beam is to change the color of the dye, so that a pattern of color changes is caused by writing data. The processing of data before writing is very much the same as for an audio disc and is covered by a specification called the Orange Book.

Reading is carried out by the same laser beam, producing reflections that depend on the color of the disc at the point being read, and detected by a photocell that will produce a strong signal for one color and a weak signal for the other color. From then on the data can be gathered and checked in much the same way as for an audio CD. The system works well, though some discs in early days had a lifetime that was much shorter than the 20−100 years that was predicted. At the time, some CD-R discs could not be played on some makes of CD players, but these incompatibilities have by now been overcome.

The success of the writable CD led inevitably to the rewritable (RW) type. The technology, however, is quite different. Though the basis is a blank plastic disc with a spiral pattern, the coating is a metal alloy whose characteristics can be changed by a low-power laser beam. Several metal alloys can exist in two forms, one as crystals and the other as a shapeless (amorphous) powder, and heating can cause the change to come about. For a suitable alloy, the two forms must have very different light reflection so that a reading laser beam will be reflected much more from one form than from the other.

This scheme works, but ran into difficulties originally because the difference in reflected beam strengths was not of great as for the CD-R, so that early types of computer optical drives (as well as most audio CD players) could not read CD-RW discs. Modern optical drives can cope with CD-R and CD-RW equally easily. DVD drives inevitably followed, providing a data space of around 4.7 Gbytes for a single-layer disc, and around double for a dual-layer disc (using layers of different depths that are written or read using a change in laser focusing). Home users found in general that the double layer was not ideal in terms of convenience because there had to be a break between recording on one layer and switching to the other. For large amounts of data storage, therefore, the later Blu-ray system is preferred to the double-layer DVD.

Digital Compression

The CD system was devised in the late 1970s, and though many aspects of the system had been tried out in the Philips video discs earlier, the technology of the 1970s was stretched to achieve the results that have already been quoted. It was not until later, with the growth of computer systems, that the remarkable effect of compression of digital signals could be achieved, making the CD system seem by modern standards overengineered.

Definition

Unlike an analog signal, a digital signal can be compressed. What this amounts to is that each sample of an analog signal is unlikely to need a complete set of bits for a digital representation, and it is also unnecessary to repeat identical codings when several successive samples are identical. Compression systems can greatly reduce the number of bits in a digital signal, and the bandwidth needed to carry them.

Definition

A compression system can be **lossy** or **non-lossy**. A non-lossy system allows the compression to be exactly reversed, so that the decompressed (expanded) signal is identical to the signal before compression. A lossy system cannot be reversed to provide the exact original, but what is lost may be unnoticeable, and the potential for compression is very much greater.

When an analog signal is converted to digital form it needs more bandwidth, as we have seen, but the number of digits that need to be used can be greatly reduced by compression methods that cannot be applied to an analog signal. Take a simple visual example, a picture that contains some sky, some land, and some sea. We may convert this picture into signals by scanning it on an 800×600 dot matrix, giving a total of 480,000 dots. If we use 16 bits to represent the brightness and color of each dot then the total number of bits for the picture is 2,880,000 bits, or 2.74 megabits (since 1 Mbit $= 102 \times 1024$ bits). This type of representation of a picture is called a **bit-map**.

Suppose, however, that all the sky is light blue, all the land is green, and all the sea is dark blue. I know it's unlikely, but please bear with me. This allows the digital information to be greatly compressed. The sky will need 16 bits as a code for the light blue color, followed by the number of identical sky-colored dots. The ground will need one 16-bit number for green, followed by the number of green-code dots. The sea will need a 16-bit number for dark blue, then the number of these dots.

Though this example is a gross oversimplification, it illustrates by how much the picture information can be reduced. This example is a non-lossy compression method called run-length encoding (RLE). More elaborate versions of RLE can be devised that find repetitive pieces of data and replace all but one of these with a location code. Typical of such non-lossy systems are **Huffman** and **LZW** coding.

▪ Note

Non-lossy methods can be used for picture files and other data that must be recovered intact, but much more can be achieved by using lossy compression methods. In practice, picture data can be compressed by up to 48-fold using a system called JPEG; the initials stand for Joint Photographic Experts Group, the committee that drew up the standard. This type of compression is used in digital still cameras so that the amount of memory needed to store images is not too great.

▪

We can also use other forms of compression. Many of the samples we take will not need the full set (16 bits in the example) of digital bits, so compression can be achieved by **bit reduction**, for example, omitting bits that are always a set of 0s. This is more complicated, but it helps to achieve very impressive compression for sound signals. Yet another form of compression can be used for video pictures, making use of the fact that in a set of successive frames of a typical video, very few parts of the picture change. For example, the background may remain unchanged while a human figure moves. In this example, the details of the background can be converted to digital form once, and the following data concern the moving body alone, not repeating the background.

None of these forms of data compression can be useful, however, unless some standardization is achieved. Systems for data compression will always make use of several types of methods, adapting the type and amount of compression to the data so that the optimum compression is achieved. This type of adaptive system is at the heart of digital television and audio broadcasting, and it is fortunate that the complications are 'buried in silicon'; in other words, the system has been worked out and the chips have been manufactured, so that the designer does not have to work out the details from scratch.

Summary

Digital coding for sound or pictures has the advantage over analog coding that the amount of information can be compressed. Basically this is done by eliminating repetition and by omitting unnecessary digits. These systems are useful only if they are standardized.

MPEG

When the CD system was launched no form of compression of data was used. The whole system was designed with a view to recording an hour of music, using 16-bit coding, on a disc of reasonable size, and the laser scanning system that was developed was quite capable of achieving tight packing of data, sufficient for the needs of audio.

Data compression was, by that time, fairly well developed, but only for computer data, and by the start of the 1980s several systems were in use. Any form of compression had to be standardized so that it would be as universal as the compact cassette and the CD, and in 1987 the standardizing institutes started to work on a project known as EUREKA, with the aim of developing an **algorithm** (a procedure for manipulating data) for video and audio compression. This has become the standard known as ISO MPEG Audio Layer III. The letters MPEG mean Motion Picture Expert Group, because the main aim of the project was to find a way of tightly compressing digital that could eventually allow a moving picture to be contained in a compact disc, even though the CD as used for audio was not of adequate capacity.

As far as audio signals are concerned, the standard CD system uses 16-bit samples that are recorded at a sampling rate of more than twice the actual audio bandwidth, typically 44 kHz. Without any compression, this requires about 8.8 Mbytes of data per minute of playing time. The MPEG coding system for audio allows this to be compressed by a factor of 12, without losing perceptible sound quality. If a small reduction in quality is allowable then factors of 24 or more can be used. Even with such high compression ratios the sound quality is still better than can be achieved by reducing either the sampling rate or the number of bits per sample. This is because MPEG operates by what are termed **perceptual coding techniques**, meaning that the system is based on how the human ear perceives sound.

The MPEG system for digital audio is based on removing data relating to frequencies that the human ear cannot cope with. Taking away sounds that you cannot hear will greatly reduce the amount of data required, but the system is **lossy**, in the sense that the removed data cannot be reinstated. The compression systems that are used for computer programs, by contrast, must not be lossy because every data bit is important; there are no unperceived data. However, compressing other computer data, notably pictures, can make use of very lossy methods, so the JPEG form of compression can achieve even higher compression ratios.

The two features of human hearing that MPEG exploits are its **non-linearity** and the **adaptive threshold of hearing**. The threshold of hearing is defined as the level below which a sound is not heard. This is not a fixed level; it depends on the frequency of the sound and varies even more from one person to another. Maximum sensitivity occurs in the frequency range 2–5 kHz. Whether or not you hear a sound therefore depends on the frequency of the sound and the amplitude of the sound relative to the threshold level for that frequency.

The masking effect is particularly important in orchestral music recording. When a full orchestra plays at maximum level, the instruments that contribute least to the sound are, according to this theory, not heard. A CD recording will contain all of this information, even if a large part of it is redundant because it cannot be perceived. By recording only what can be perceived, the amount of music that can be recorded on a medium such as a CD is greatly increased, and this can be done without any perceptible loss of audio quality for the *average* listener. A musically trained ear can hear the differences, particularly in classical music (though oddly enough musicians may put up with poor-quality recordings, perhaps because their brains are programmed to hear what was intended).

MPEG coding starts with circuitry described as a **perceptual sub-band audio encoder** (PSAE). The action of this section is to analyze continually the input audio signal and from this information prepare data (the masking curve) that define the threshold level below which nothing will be heard. The input is then divided off in frequency bands, called **sub-bands**. Each sub-band is separately quantized, and data on the quantization used for a sub-band is held along with the coded audio for that sub-band, so that the decoder can reverse the process. Figure 12.5 illustrates a simplified block diagram for this process.

Summary

MPEG is a set of standardized methods for compressing sound and picture data, allowing more information to be stored on a disc, and also permitting the use of digital radio and digital television.

MPEG Layers

Definition

MPEG 2, as applied to audio signals, can be used in three modes, called layers I, II, and III. Ascending layer numbers mean more compression and more complex encoding. Layer I is used in home recording systems and for solid-state audio (sound that has been recorded on chip memory, used for automated voices, etc.). Equipment that works on a given layer will work on lower levels, but not on higher levels. For example, a layer II decoder will work with layer I signals.

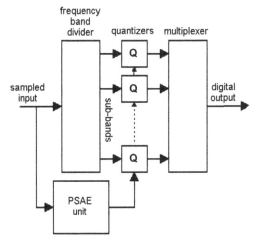

Figure 12.5:
A block diagram of the MPEG audio compression system

Layer II offers more compression than layer I and is used for digital audio broadcasting, television, telecommunications, and multimedia work. The bit rates that can be used range from 32 to 192 kb/s for mono, and from 64 to 384 kb/s for stereo. The highest quality, approaching CD levels, is obtained using about 192–256 kb/s per stereo pair of channels. The precise figure depends on how complex an encoder is used. In general, the encoder is from two to four times more complex than the layer I encoder, but the decoder need be only about 25% more complex. MPEG layer II is used in applications such as CD-i full motion video, video CD, solid-state audio, disc storage and editing, digital audio broadcasting (DAB), DVD, cable and satellite radio, cable and satellite television, integrated services digital network (ISDN) links, and film sound tracks.

Layer III offers even more compression, and is used for the most demanding applications for narrow band telecommunications and other specialized professional areas of audio work. It has found much more use as a compression system for MP3 files.

■ Note

MPEG-2 is one of several (seven at the last count) MPEG standards, and we seem to be in danger of being buried under the weight of standards at a time when development is so rapid that each standard becomes out of date almost as soon as it has been adopted. Think, for example, how soon NICAM has become upstaged by digital television sound. More confusion is caused by the use of numbers that are not in sequence (so that MPEG-21 does not mean that there are MPEGs 20, 19, … and so on). The current standard for digital television and sound broadcasting is MPEG-2, and at one time a variation (MPEG-3) was suggested for high-definition television, but this has not been implemented because of changes in the ways that MPEG-2 could be compressed.

MP3 Audio

Definition

MP3 is a high-compression digital sound coding and decoding system that is now used for transmitting audio signals over Internet links and for storing audio signals in compact computer file form. MP3 allows the construction of small players that store, typically, several hours of music, but contain no moving parts.

■ Note

Because MP3 is a lossy form of compression the MP3 deck for hi-fi systems has not emerged so far, but we should remember that the compact cassette was also considered unfit for hi-fi use in its initial days.

The name MP3 began as an extension to a computer filename, devised to distinguish sound files created using MPEG-1 layer III encoding and decoding software. The PC type of computer makes use of these extension letters, up to three of them, placed following a dot and used to distinguish file types. For example, **thoughts.txt** would be a file called thoughts, consisting purely of text, and **thoughts.doc** would be a document called thoughts which could contain illustrations and formatted text, even sounds. A file called **thoughts.jpg** would be a compressed image file and **thoughts.bmp** would be an uncompressed image file. There are many such extensions, each used to identify a specific type of file.

■ Note

The same mp3 extension is used for sound files that have used MPEG-2 layer III with a reduced sampling rate, but there is no connection between MP3 and MPEG-3. MP3 files use a compression ratio of around 12:1, so that MP3 files stored on a recordable CD (an MP3-CD) will provide about nine hours of sound. Older CD players will not work with such a disc, but the more modern units will (they are distinguished by the MP3 logo on the casing).

The main use of MP3, however, has been the portable MP3 player, which allows MP3 files to be recorded from downloads over the Internet. This started off as very much of an audio system for the computer buff, but like all matters pertaining to computing, this use has spread. MP3 is unlikely to appeal to those who seek perfection in orchestral music (let's face it, what system does?), but for many other applications it offers a sound quality that is at least as good as anything that can be transmitted by AM radio or obtained from a cassette.

The advantages are many. You can load the memory up with music that you like, deleting anything you do not want to hear again. You can play tracks in any order, select tracks at random, and store other music on your PC until you want it on your MP3 player.

Summary

MP3 uses a lossy compression, typically around 12:1, on sound files to allow for transmitting such files over the Internet, storing them in solid-state memory, magnetic drives or CD format, and replaying in small units that contain no moving parts. This type of system is the successor to the compact cassette.

CHAPTER 13

Microprocessors, Calculators, and Computers

The Microprocessor

Definition

A **microprocessor** is a single logic chip that contains gates and flip-flops which are arranged so that the connections between the internal units are controllable. The control is arranged by using a set of bits that form a **program** input to the microprocessor and which are stored in a register (the **program register**). Another form of processor, the microcontroller, is a more complex device that combines the actions of a microprocessor with circuitry that carries out actions such as counters, timers, communications, and analog inputs so that it can be used as a complete control device on a single chip.

The microprocessor can **address** memory, which means that it can select an item of stored data from any location in external memory chips and make use of it, and it can store results also in a selected portion of the memory. Within the microprocessor chip itself, logic actions such as the standard NOT, AND, OR, and XOR gate actions can be carried out on a set of bits, as well as a range of other register actions such as shift and rotate, and some simple arithmetic. The important item that provides the definition of a microprocessor is that any sequence of such actions can be carried out under the control of a program which is also read in from memory.

Figure 13.1 shows a block diagram that is extensively used for teaching purposes, and which is still applicable to many types of microprocessor that are used for industrial control actions. Before we can look at the action we need to know what the units are and how they are connected. One important point is the use of **bus** connections.

Definition

A bus is a set of conducting lines that connects several units, each line being used for a single-bit signal. Most buses are bidirectional, meaning that signals can be sent in either direction, and this is achieved by using enabling pulses to the units that are exchanging bits.

For example, if a unit is enabled and its outputs are connected to the lines of a bus, then a set of bits will be placed on the bus lines and these bits can be used as inputs for any other unit whose

Electronics Simplified. DOI: 10.1016/B978-0-08-097063-9.10013-5

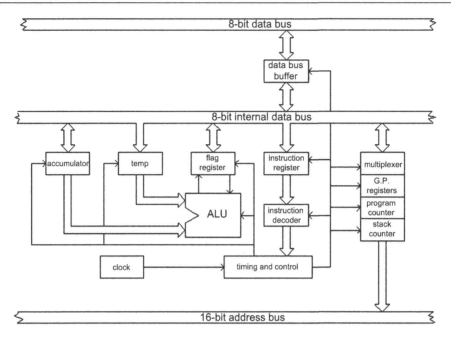

Figure 13.1:
A block diagram of the circuits inside an older type of 8-bit microprocessor. Modern chips are much more elaborate, but the basic principles are the same

inputs are connected to the bus and which is also enabled. In practice, both the inputs and the outputs of each unit will be connected to the same bus, and the enabling pulses allow reading (input), writing (output), or disabled (not connected) states. The use of buses allows the construction of the microprocessor to be much simpler than it would be if separate connections had to be used for each set of inputs and outputs.

In use, the main data bus can be connected with other units, such as memory and latches, to allow the microprocessor to make use of these other circuits. The two important buses that are used in this way are the **data bus** and the **address bus**, and these, along with control signals on a **control bus**, allow the microprocessor to control other units as if they were a part of the microprocessor itself.

Summary

A microprocessor is a type of universal logic integrated circuit (IC) chip which can carry out a set of actions in sequence. The sequence and the actions are controlled by a program, which consists of binary numbers that are used to open and close gates within the microprocessor. Other units (such as memory) are connected by way of buses, sets of lines which connect to all units. When a bus is used to pass signals, the sending and the receiving units are enabled, and other units are disabled, allowing the same buses to be used for different signals in either direction.

Microprocessor Cycle

Definition

All microprocessors work on a **cycle**, meaning a fixed routine that involves reading from memory, carrying out an action, and writing to memory. The sequence of actions is maintained so as to avoid confusion between a program byte and a data byte.

The action, simplified, is as follows. When the microprocessor starts work, it will read a byte from some fixed address in the external memory store. This first byte is *always* an instruction, and the microprocessor reads it into an internal register, the **instruction register**, where it is decoded. As a result of reading the instruction code (by matching it with one of a set permanently stored in the microprocessor), an action will be carried out. This action is controlled by a short built-in program, called a **microprogram**.

Definition

A microprogram is a short routine for setting the connections within the microprocessor to carry out an action, such as ANDing two bytes. There is a microprogram stored in the microprocessor for each possible microprocessor action. Each step of a microprogram will set gates or activate registers inside the microprocessor. The stored microprograms determine what the microprocessor can carry out.

This microprogram action may require data; for example, if you want to AND two bytes together then these bytes constitute the data, so that a program command for ANDing two bytes would have to be followed by reading the two bytes that were to be ANDed.

■ Note

More usually, one byte will already be stored in a **register**, usually the **accumulator** register of the microprocessor, and only one other byte will need to be read from memory.

■

The instruction code must contain the information that allows the microprocessor to start a microprogram, then find and read the data (in a register or in the memory). When the data has been read the action will be carried out and the result is stored in a register called the **accumulator** (or it can be stored in memory if another instruction provides for this).

The next byte that the microprocessor reads will then be another instruction. The only difference between a byte of instruction and a byte of data is the way that they are arranged in the memory, instruction first, and its data following, and this correct sequence ensures that the bytes are routed to the correct registers in the microprocessor. The program must be written so that this sequence is always obeyed. The timing of actions is determined by clock pulses which are supplied by a crystal-controlled oscillator, usually at a high speed (as low as 400 MHz for a little notepad computer but close to 3 GHz for a desktop or a fully specified laptop computer). Each action usually requires more than one clock pulse because of the number of steps in an action, even a simple action such as adding.

All actions such as AND or ADD are carried out in the arithmetic and logic unit (ALU), and this is arranged so that one of its inputs is from the accumulator register and its output also is to the same register; remember that the microprocessor is clock controlled, so that input and output occur at different times. This arrangement means that the accumulator always provides one of the bytes in any arithmetic or logic action, and it always holds the result following such an action. If a second input is required, as it is for most actions, it can be supplied from the temporary register which in turn is filled from any unit connected to the internal data bus. The advantage of using registers is that the time needed to fetch a byte from a register (or to write a byte to a register) is much less than would be needed for reading or writing memory.

■ Note

The 'data size' of a microprocessor is stated in terms of the number of bits in the accumulator register. Early microprocessors (still in use for industrial applications) had an 8-bit register and are therefore 8-bit microprocessors. By 1982, computers using 16-bit microprocessors were on sale. More recent types are 32-bit and 64-bit microprocessors. In these later processors several registers can be used as accumulators, and some processors can contain several **cores**, meaning that more than one action can be carried out at the same time; this is like having more than one microprocessor in one casing.

Summary

All computing actions start by reading a program code from memory. This code is placed in the program register of the microprocessor and used to activate a microprogram that is stored within the microprocessor itself. This may require further number data to be read from memory, and the programmer must ensure that these numbers follow the program code(s). All microprocessors make use of an **accumulator** register that can supply one number for an action, and where the result of an action will also be stored. The size of the accumulator in terms of bits is used as a measure of the data unit capability of the microprocessor (as 8-bit, 16-bit, 32-bit, 64-bit, and so on).

Memory

Definition

Memory is a name for a type of digital circuit component which can be of several varieties. The basis of a single unit of memory is that it should retain a 0 or a 1 signal for as long as is required, and that this signal can be connected to external lines when needed for reading (copying) or writing (storing).

The method that is used to enable connection is called **addressing**. The principle is that each unit of memory should be activated with a unique combination of signals that is present on a set of lines called the **address bus**. Since each combination of bits constitutes a binary number, the combinations are called **address numbers**. We have already looked at this idea in connection with encoders and multiplexers.

Definition

There are two fundamental types of memory, both of which are needed in virtually any microprocessor circuit. One type of memory contains fixed unalterable bits, and is therefore called read-only memory, or **ROM**. The important feature of ROM is that it is **non-volatile**, meaning that the stored bits are unaffected by switching off power to the memory and they are available for use whenever power is restored. For example, the microprograms are stored in ROM that is contained within the microprocessor itself.

Since there must be an input to the microprocessor whenever it is switched on, ROM is essential to any microprocessor application, and in some applications it may be the only type of memory (outside the microprocessor itself) that is needed. The simplest type of such a ROM consists of permanent connections to logic 0 or logic 1 voltage lines within a chip, gated to output pins when the correct address number is applied to the address pins.

Definition

Read–write memory is the other form of memory, which for historical reasons is always known as **random-access memory** (**RAM**). This is because, in the early days, the easiest type of read–write memory to manufacture consisted of a set of serial registers, from which bits could be read at each clock pulse in sequence. This meant that if you wanted the 725th byte you had to read, but not use, the first 724. Random access means that you can read any byte without needing to read others.

The use of addressing means that any byte can be selected at random, without having to read out all the preceding bytes; hence the name 'random access'. Practically all forms of memory that are used nowadays in microprocessor systems feature random-access addressing, but the name has stuck as a term for read—write memory. Memory of this type needs address pins, data pins, and control pins to determine when the chip is enabled and whether it is to be written or read. The reading of a memory of this type does not alter the contents.

Summary

A microprocessor must be used in conjunction with memory which is used to hold program and data bytes and also for retaining the results of actions carried out by the microprocessor. Permanent (read-only) memory (ROM) is needed to hold instructions that need never change (allowing, for example, the microprocessor to be used with a keyboard and a display of some kind). Read—write memory, called RAM, is needed so that it can be loaded with program and data bytes (usually from magnetic discs) that the microprocessor can work on. Only electronic memory operates rapidly enough to allow the microprocessor to run at its intended speed.

■ Note

Unlike ROM, RAM is normally volatile; its contents are lost when the supply voltage to the chip is switched off. It is possible, however, to fabricate memory using complementary metal-oxide superconductor (CMOS) techniques and retain the data for very long periods, particularly if a low-voltage backup battery can be used. Such CMOS RAM is used extensively in calculators and is used in computers to hold essential setup data, including time and date information.

Static RAM is based on a flip-flop as each storage bit element. The state of a flip-flop can remain unaltered until it is deliberately changed, or until power is switched off, and this made static RAM the first choice for manufacturers in the early days. The snag is that power consumption can be large, because each flip-flop will draw current whether it stores a 0 or a 1. This has led to static RAM being used only for comparatively small memory sizes and where very fast operation is needed.

■ Note

Static RAM is extensively used in modern computers as **cache** memory. This is very fast but small memory that is used by the computer as a temporary store. The principle is that the cache memory can read bytes from the (slow) main memory at times when the processor is working without using memory, and the processor will then read data from the cache, which is a much faster action than reading from the main memory. The same applies to writing to memory: the microprocessor writes quickly to the cache, and the cache writes at a leisurely pace to the main memory. This process is not infallible because the cache may not contain the correct data, forcing the microprocessor to read from the main memory, but it is not unusual to find that the cache is used for 90% of the read actions. A suitable program has to be used to control the action of the cache writing and reading.

■

Dynamic RAM

The predominant type of RAM technology for large memory sizes is the **dynamic RAM**. Each cell in this type of RAM consists of a miniature metal-oxide superconductor (MOS) capacitor with logic 0 represented by a discharged capacitor and logic 1 by a charged capacitor. Since each element can be very small, it is possible to construct very large RAM memory chips (128 Mb × 4-bit is now common), and the power requirements of the capacitor are very small. A typical RAM memory card nowadays is 1 GB or 2 GB (subject to limitations set by the operating system).

The snag is that a small MOS capacitor will not retain charge for much longer than a millisecond, since the insulation of the capacitor cannot be perfect and will inevitably allow leakage. All dynamic memory chips must therefore be **refreshed**, meaning that each address which contains a logic 1 must be recharged at intervals of no more than a millisecond. The refreshing action can be carried out automatically within the chip and the user is never aware of the action. The very large memory chips used in modern computers are all of the dynamic type.

■ **Note**

The capacity of a memory is given in terms of megabytes (Mbyte) or gigabytes (Gbyte) (1 Mbyte = 1024 Kbyte, or 1,0485,76 bits, and 1 GB is 1024 MB). Static RAM for fast cache memory is typically used in units of around 64–256 Kbyte. Dynamic RAM is used for the main memory, typically 1–8 Gbyte in modern computers.

■

Summary

RAM is volatile, meaning that it loses its data when the power supply is switched off. Static RAM is expensive but very fast, and it is used in small quantities in modern computers. The main memory invariably uses dynamic RAM minicapacitors which lose data in a short time, a few milliseconds, but which can be refreshed automatically. The low cost of dynamic RAM allows large memories to be used, typically 1 Gbyte upwards.

Buses

The buses of a microprocessor system, as introduced earlier, consist of lines that are connected to each and every part of the system, so that signals are made available at many chips simultaneously, and can be passed between any pair of chips. The three main buses are the address bus, the data bus, and the control bus. Since understanding the bus action is vitally important to understanding the action of any microprocessor system, we will concentrate on each bus in turn, starting with the address bus.

Definition

An address bus consists of the lines that connect between the microprocessor address pins and the address pins of each of the memory chips in the microprocessor system. In anything but a very simple system, the address bus would connect to other units also, but for the moment we will ignore these other connections.

A typical older-style 8-bit microprocessor would use 16 address pins. Using the relationship that n pins allow 2^n binary number combinations, the use of 16 address lines permits 65,536 memory addresses to be used, and modern computing microprocessors use 20, 24, 32, or 64 address lines, of which the use of 64 lines is the most common now.

Most early types of memory chips were 1-bit types, which allowed only 1 bit of data to be stored per address. For a 64K 8-bit microprocessor, then, the simplest RAM layout would consist of eight 64K $\times$ 1-bit chips, each of which would be connected to all 16 lines of the address bus. Each of these chips would then contribute 1 bit of data, so that each chip was connected to a different line of the data bus. This scheme is illustrated in Figure 13.2.

■ Note

For modern computers, memory is not installed in single chips. Chips are assembled into units called dual in-line memory modules (DIMMs), which use a standard plug-in connection. Because of rapid development in memory construction, these DIMM units come in so many versions that great care is needed in selection of memory, and if you want to upgrade the memory in a computer you should consult the tables

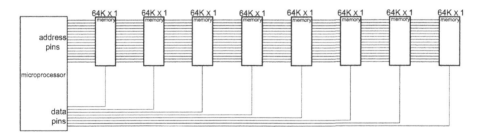

Figure 13.2:
How 64K of memory is arranged for a simple 8-bit computer. Modern machines can use much larger amounts of memory, but the principles are similar

provided by distributors such as Crucial or Kingston. At the time of writing, the type of DIMM in current use is labeled DDR3 and the DIMM board needs a 240-pin holder.

At each of the 65,536 possible address numbers of a 16-bit system, each memory chip will give access to 1 bit, and this access is provided through the lines of the data bus. The combination of address bus and data bus provides for addressing and the flow of data, but another line is needed to determine the direction of data.

This extra line is the **read/write line**, one of the lines of the **control bus** (some microprocessors use separate read and write outputs). When the read/write line is at one logic level, the signal at each memory chip enables all connections to the inputs of the memory units, so that the memory is written with whatever bits are present on the data lines. If the read/write signal changes to the opposite logic level, then the internal gating in the memory chips connects to the output of each memory cell rather than to the input, making the logic level of the cell affect the data line (placing bits on the data lines). In addition, there is usually one or more enable/disable lines so that the memory can be disabled when addressing is used for other purposes.

Summary

Memory chips are connected to the bus lines, and since it is quite common for a chip to store in 1-bit units, one memory chip may be needed for each bit of a full byte. On modern computers, the chips are assembled into DIMM units which can be plugged easily into the computer. In addition to the address and data lines, the memory must use read/write signals to determine the direction of data flow, and enable/disable signals to allow the whole memory to be isolated when the address bus is being used for other purposes.

■ Note

Dynamic memory chips use, in practice, a rather different addressing system: each address consists of a column number and a row number. This is done to make refreshing simpler, and the address numbers on the address bus need to be changed into this format by a memory manager chip. This does not affect the validity of the description of memory use in this chapter.

The provision of address bus, data bus and read/write lines will therefore be sufficient to allow the older type of 8-bit microprocessor to work with 64 Kbyte of memory in this example. For smaller amounts of memory, the only change to this scheme is that some of the address lines of the address bus are not used. These unused lines must be the higher order lines, starting at the most significant line. For a 16-line address bus, the most significant line is designated as A15, the least significant as A0.

A memory system for an 8-bit processor that consisted purely of 64K of RAM, however, would not be useful, because no program would be present at switch-on to operate the microprocessor. There must be some ROM present, even if it is a comparatively small quantity. For some control applications, the whole of the programming might use only ROM, and the system would consist of one ROM chip connected to all of the data bus lines, and as many of the address lines as were needed to address the chip fully. As an example, Figure 13.3 shows what would be needed in this case, using an 8K × 8-bit ROM, which needs only the bottom 13 address lines.

It is more realistic to assume that a system will need both ROM and RAM, and we now have to look at how these different sets of memory can be addressed. In the early days, the total addressing capability of an 8-bit machine was no particular restriction, and a common configuration was of 16K ROM and 16K RAM. This could be achieved by 'mapping the memory' as shown in Figure 13.4; other combinations are, of course, possible. In the scheme that is illustrated, the ROM uses the first 16K of addresses, and the RAM uses the next 16K. Now the important thing about this scheme is that 16K corresponds to 14 lines of an address bus, and these same 14 lines are used for **both** sets of memory.

The lower 14 address lines, A0 to A13, are connected to both sets of chips, represented here by single blocks. Line A14, however, is connected to chip-enable pins which, as the name suggests, enable or disable the chips. During the first 16K of addresses, line A14 is low, so that ROM is enabled (imagining the enable pin as being active when low) and RAM is disabled. For the next 16K of addresses on lines A0 to A13, line A14 is high, so that ROM is disabled and RAM is enabled. This allows the same 14 address lines to carry out the addressing of both ROM and RAM. A simple scheme like this is possible only when both ROM and RAM occupy the same amount of memory and require the same number of address lines.

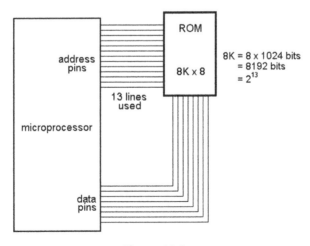

Figure 13.3:
Connecting a simple microprocessor to an 8K ROM chip

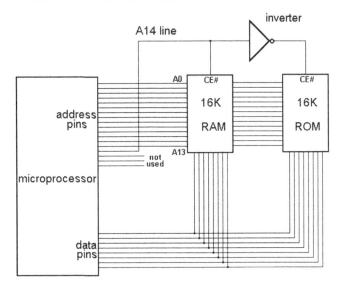

Figure 13.4:
Connecting 16K of RAM and 16K of ROM so that each memory unit can use different address numbers

Summary

The same address lines are used for both ROM and RAM chips, using enable/disable pins to ensure that one range of address numbers will activate ROM and another range will activate RAM. Though 8-bit examples have been shown here, the principles apply (but with much more elaborate drawings) to modern 64-bit systems.

PC Register and Addressing

Definition

The microprocessor runs a program by outputting address numbers on the address bus so as to select memory. At each memory address, one or more bytes will be read to obtain an instruction or the data to carry out an instruction, or bytes can be written to store in memory.

The sequence of reading memory is normally a simple incrementing order, so that a program that starts at address 0000 will step to 0001, 0002, 0003, and so on, automatically as each part of the program is executed. The exception is in the case of a **jump**, caused by an **interrupt** (see later) or by a software instruction. A jump in this sense means that a new address will be placed into the program counter register, and the microprocessor will then read a new instruction starting at this address and incrementing the address number at the end of each instruction. For the moment, however, the important point is that the normal action is one of incrementing the memory address each time a program action has been executed, and the sequence of actions is all important.

■ Note

Many microprocessors are designed so that they will automatically read a byte from some fixed address when they start operating. This byte is always a jump instruction, followed by data, that will cause another address range to be used. This allows the programmer to specify where the main program will be located.

■ Note

The **program counter** (**PC**) or **instruction pointer** (**IP**) register is the main addressing register, and is connected by gates to the address pins of the microprocessor. The number in this register will be initialized by a voltage applied to a RESET pin, and it is automatically incremented each time an instruction has been executed, or when an instruction calls for another byte.

Imagine, for example, that the whole of the RAM memory from address 0000 is filled with an NOP instruction byte. NOP means no operation, and its action is simply to do nothing, just go on to the next instruction. If the PC is reset to contain the address 0000, then the NOP byte (which might be 00) at this address will be read, decoded, and acted on. The action is nil, and so the PC is incremented to address 0001, the byte at this address is read, and the action (or lack of action) is repeated. Figure 13.5 shows a simplified timing diagram for the Z80 processor, indicating what actions are involved in reading an instruction byte; note that several clock cycles are needed to carry out one simple instruction.

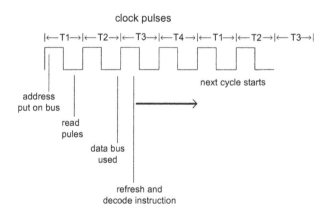

Figure 13.5:
Clock pulse and actions. An outline of the stages in carrying out a simple instruction

If the entire memory is filled in this way, the microprocessor will simply cycle through all of the memory addresses until the address reaches 0000 again, and the whole addressing sequence will repeat. Of course, in a real-life system, the memory is not full of NOP bytes. The timing and the PC actions then depend very much on what instruction bytes are present, and even more so on the addressing method.

■ **Note**

The Z80 is an old microprocessor, no longer used for computers, but it is extensively used in industrial control applications. Its operation forms an excellent introduction to microprocessor programming.

■

Looking at addressing methods brings us into the realm of software, but it is necessary for understanding how the PC and buses can be used during an instruction that involves the use of memory. The principle is simple enough: that many of the instructions of the microprocessor require a byte (or more) to be obtained from the memory.

Carrying out instructions like NOP, or the register shift and rotate instructions, does not normally require any other load from memory. This is because these actions are carried out on a single byte, word (16 bits), or Dword (32 bits), which can be stored in one of the registers of the microprocessor. For a lot of actions, though, one word will be stored in a register, and another word must be taken from memory. This is controlled by programming: the instruction byte or word is followed by a set of bytes that is the address number for where the data is stored, or some number that allows the address to be calculated.

Summary

The microprocessor operates in sequence, and when power is applied, it will read a byte from some specified location, which may be the first memory byte but is often arranged to be a byte near the last possible memory address. This first byte will often be a jump instruction that causes the processor to read another byte at another location, so allowing a programmer to decide where the main program is placed. From then on, bytes are read in sequence unless a jump command causes a change to another memory location. The program counter (PC) or instruction pointer (IP) register is used to store each address, and unless a jump is encountered, this register is incremented after each instruction.

Interrupts

Definition

An **interrupt** is a signal that interrupts the normal action of the microprocessor and forces it to do something else, almost always a routine which starts at a different address and which will carry out an action that deals with the needs of the interrupt signal. Such a routine is called an

interrupt service routine. Following this routine, the microprocessor must resume what it was doing, so that memory must be used to store numbers that will allow resumption. A typical example is the use of a keyboard: when a key is struck an interrupt routine swings into action to read which key has been pressed and to store a code in memory.

The interrupt is an electrical signal or a software instruction. An electrical pulse can be applied to one of the interrupt pins of the microprocessor; it can come, for example. from an external switch, or a software instruction can be read to force an interrupt. When this signal is received, the microprocessor executes an interrupt routine. In doing so, it will complete the instruction that it is processing, store the numbers it is using, and then jump to an address to get directions for a service routine, in this example the routine that reads the keyboard.

■ **Note**

When the service routine is completed, the microprocessor resumes its program instructions (Figure 13.6). All microprocessors allow a part of the memory to be designated as a **stack**. This means simply that some addresses are used by the microprocessor for temporarily storing register contents, making use of the memory in a very simple last-in-first-out way.

■

Precisely which addresses are used in this way is generally a choice for the software designer. When an interrupt is received, the first part of the action is for the microprocessor to complete the action on which it is engaged. The next item is to store the PC address in the stack memory. This action is completely automatic. Only the address is stored in this way, however. If the

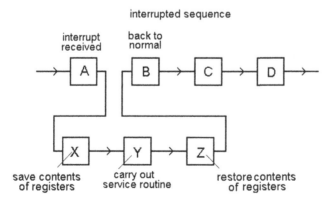

Figure 13.6:
How an interrupt works. While the interrupt routine is activated, the state of all registers must be stored in memory so that the microprocessor can recover its register contents when the main program is resumed

interrupt service routine will change the contents of any other registers of the microprocessor, it will be necessary to save the contents of these registers on the stack also.

This is something that has to be attended to by the programmer who writes the interrupt service routine, saving the register contents at the start of the interrupt service routine and replacing them afterwards. Finally, the problem of what happens if an interrupt signal arrives while another interrupt is being serviced is dealt with by disabling the interrupt mechanism while an interrupt is being serviced. This is not necessarily automatic, and will often form one of the first items in the interrupt service routine. The tendency over the short history of microprocessors has been to delegate these actions to the software writer rather than to embed them in hardware.

Summary

The automatic action of reading bytes in succession can be interrupted by a signal to an interrupt pin or a software interrupt command. This action will cause the microprocessor to start taking instructions (the service routine) from another location in the memory, and when this program (a subroutine) is completed, the previous action will be restored. Essential data are stored in part of the memory, called the stack, when an interrupt is called, and are restored when the interrupt is completed.

Inputs and Outputs

A system that consisted of nothing more than a microprocessor and memory could not do much more than mumble to itself. Every useful system must have some method of passing bytes out of the system to external devices, and also into the system from external devices. For computer use, this means at the very least the use of a visual display unit (VDU) and a keyboard, however primitive.

For modern systems, you can add to this disc drives, floppy (now obsolete) or hard, CD- or DVD-ROM, various sockets into which other add-ons can be plugged (such as printer, scanner, light pen, joystick, mouse, trackerball, etc.), and possibly A-to-D inputs for measuring voltages. If the microprocessor system is intended for machine-control purposes, then the outputs will be very important indeed, because these will be the signals that will control the machine. Figure 13.7 shows a block diagram for a computer system using a microprocessor.

The inputs may be instruction codes from a disc, pulses from limit switches, digital voltage readings from measuring instruments, and so on. Whatever the function of the system, then, the inputs and outputs form a very important part of it all. A surprising number of control actions of a microprocessor system, in fact, consist of little more than passing an input signal to an output device, perhaps with some monitoring or comparison action thrown in. The action of passing

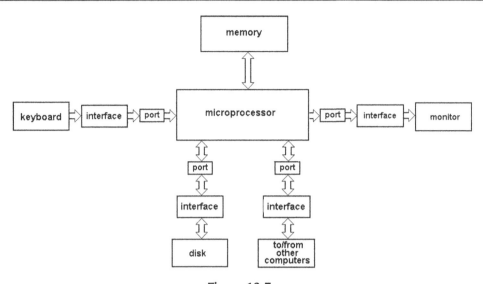

Figure 13.7:
A block diagram for a computer, omitting details

signals between a microprocessor and other units is called **interfacing**, and it is usually the most difficult part of making use of a microprocessor.

■ Note

Before we get involved with the details of how signals are passed between a microprocessor system and the rest of the world, we need to be aware of the problems that are involved. The main problem is one of timing. The microprocessor system works rapidly, governed by the rate of the system clock. There is no point in having, in the software, instructions that will make the microprocessor send a word out of the system unless you can be sure that whatever you are sending the word to can deal with it at that time.

■

The time that is involved might be perhaps a couple of clock cycles, a fraction of a microsecond, and there are not many systems, apart from another microprocessor, that can deal with such short-duration signals. The same problem applies to incoming signals. At the instant when you press a key on a keyboard, can you be certain that a microprocessor is executing an instruction that will read the key? Many such problems are dealt with by using the interrupt system, so that pressing a key on the keyboard will generate an interrupt that will activate a reading routine.

Another matter is the nature of the signals. The signals from a microprocessor system will be digital signals, typically using standard voltages of 0 V and +5 V. If the input from another system happens to be an analog signal of maximum amplitude 50 mV, or if the signal that is required by a device at the output is a +50 V, 500 Hz AC signal, then a lot of thought will have to be devoted to interfacing. This is outside the scope of this chapter, because for machine-control applications in particular, interfacing is by far the most difficult action to achieve in a new design.

Summary

A microprocessor and its memory becomes a computer system when it is connected by interfacing chips to a keyboard, display unit, disc drive(s), and power supply. The signals may have to be converted to other formats, or their timing altered, to be usable by these other units. Computing action is possible only when software programs are loaded and run.

Ports

Definition

A **port** is a circuit that controls the transfer of signals into or out of the microprocessor system. A port is likely to include some memory (a register) for temporary storage.

If the only requirement for a system is to pass a single bit in and out, the obvious method is to use a simple latch chip, connecting with one of the data lines of the microprocessor system. Even if eight data lines are needed, a straightforward hardware latch may be sufficient, particularly if the signals are all in one direction, or if most of the signals are in one direction. Ports must be used, for example, for keyboard inputs.

The port will have an address, like memory, and it can store a small amount of data. To pass data out of the system, the microprocessor uses the port address, and puts data on the data bus. The port stores the data, and sends out a signal that it contains data. The circuits beyond the port can then read the signal as and when they are ready.

In the other direction, if an external circuit sends data to the port, the port will store the data and send an interrupt signal to the microprocessor. The interrupt service routine for the port will start running, allowing the microprocessor to address the port and read the data. When this has been done, the port will send out a signal to indicate that it is ready for more data, and the microprocessor resumes its normal actions.

Summary

Port chips can be almost as complicated as microprocessor chips, and are used to connect the microprocessor buses to the external components of a computer system. Ports are programmable chips, and they carry out all the interfacing actions that are needed to make the signals compatible.

USB

The letters mean universal serial bus, a system that is replacing all the different systems that have been used on older computers for connecting peripherals such as mice, keyboards, printers, and scanners. The system uses a standardized form of cable and connectors (see Chapter 14), though there is less standardization for the miniature connectors for digital cameras. The system provides for up to 127 devices to be connected; modern desktop computers typically use six to eight USB sockets and smaller machines feature two to four.

Like anything else connected with computers (and connection computers) the USB standard has been updated several times, while retaining compatibility with the original system. The USB-1 system was a fast two-channel serial connection that could replace older serial and parallel systems. Version 1.0 appeared in 1996 and provided a low-speed (1.5 Mb/s) connection for components such as keyboard and mouse along with a faster connection (12 Mb/s) for other peripherals. An important part of the specification was that the connection could be 'hot-swappable', meaning that the USB connector could be inserted or removed on a computer that was still running. Older connectors required the computer to be shut down before a connector was inserted or removed.

In 2001 the new version, USB-2, provided for a fast rate of 480 Mb/s, and in 2008 the USB-3 standard hiked this top speed to 4.8 Gb/s. In due course, personal computers will use only USB connectors, but at present the older connectors are still installed because many users still have older peripherals that use other connection methods (parallel or serial).

Calculators

A calculator is constructed using a microprocessor and memory, with most of the memory in ROM form to provide the routines for the more complicated processes. The arithmetic actions of the microprocessor itself are confined to addition or subtraction of two bytes, sometimes with provision for multiplication and division of single bytes. Anything more complicated must be handled by software (program instructions) and these can be read from the ROM.

Virtually all calculator actions require these additional program instructions, because the calculator uses ordinary denary numbers that can use a decimal point, but the microprocessor will handle only single bytes that represent whole binary numbers. The programs in the ROM are therefore the 'clever' part of a calculator, and the rest is well standardized, a microprocessor with a small amount of RAM (for storing answers, intermediate results and constants) and an input/output port for the keyboard and the display (Figure 13.8). A few calculators are programmable, meaning that the instructions that you would normally carry out in sequence on data can be programmed into RAM and carried out on data that you can also store in RAM.

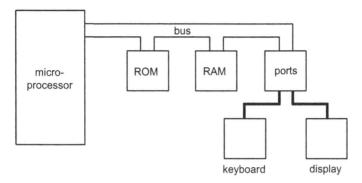

Figure 13.8:
A block diagram that is typical of any simple calculator. All units apart from keyboard and display are usually made as a single chip

▪ Note

The diagram in Figure 13.8 shows separate units, but the whole of a calculator, including microprocessor, ROM, RAM, and ports, is usually constructed as a single chip, needing only connections to the display, keyboard, switches, and battery.

▪

CHAPTER 14

Computers and Peripherals

Definition

A computer is a programmable microprocessor system with a large amount of random access memory (RAM), a keyboard and a visual display unit (VDU) as minimum requirements. Data in binary form can represent denary numbers, words, pictures, or sound. The actions of the computer are made possible by the programs that it can run. For general-purpose use a computer needs an operating system (see Chapter 15). A **peripheral** is a device that is connected to a computer to perform such actions as display, printing, selection of operations, communication, etc.

■ **Note**

The computer is just one example of a class of devices that are known as **state machines**. A state machine is anything that can store information, change information, and cause a signal or action output.

■

Computers

Figure 14.1 shows a block diagram for a generalized 8-bit data computer system which would have been typical of small computers in the early 1980s, and whose principles can still be applied to modern 32-bit machines. The block that is labeled MPU is the microprocessor unit, which will have its own internal buses.

■ **Note**

The MPU is shown as placing signals on the address bus, because in a system like this with only one processor, the MPU has total control over the address bus; all other chips **receive** address signals. The data connection between the data bus and the MPU is two-way, because the MPU must read data as an input and also output processed data. The connection of the MPU to the control bus is also two-way, because the MPU will issue some control signals to other chips and also receive signals from other chips. Note that read-only memory (ROM) sends data to the data bus but does not receive data.

■

Electronics Simplified. DOI: 10.1016/B978-0-08-097063-9.10014-7

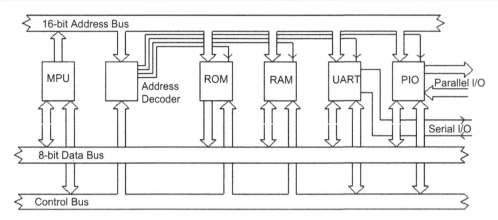

Figure 14.1:
A block diagram for an 8-bit computer, typical of types produced in the mid-1980s. Modern machines are much more elaborate, but the block diagram is typical of basic design

The address **decoder** is the chip that allocates addresses. Some MPU address numbers must correspond to storage in ROM and some to storage in RAM, and a few are reserved for the UART (universal asynchronous receiver/transmitter; serial) and PIO (programmed input/output; parallel) chips. The address decoder chip receives signals from the address bus and from the control bus (originating in the MPU) and determines which of the four memory-using sections will be enabled. This chip prevents any possibility of contention for the use of the buses, because only the chip that is entitled to use the buses will be enabled.

The **ROM** unit (usually one or two chips) receives signals from the address bus, and is enabled or disabled by an output from the address decoder. It is also controlled by control bus signals, and its output is placed on the data bus. This output will for the most part be program instructions to the MPU, such as start-up instructions, and short routines for such tasks as peripheral control (disk, VDU, keyboard, etc.).

The **RAM** unit (which may consist of a large number of individual chips, each storing one bit of data) uses the signals from the address bus, and is enabled by the address decoder. It also receives signals from the control bus, notably the read/write signals that determine the direction of data flow, and it has a bidirectional (two-way) connection with the data bus.

The **UART** chip uses the address bus signals and an enabling signal from the address decoder. It has a bidirectional connection with the data bus and also with the control bus. The control bus can determine the direction of data and can also interrupt the MPU to ensure that serial input is dealt with by a suitable routine. Two separate lines allow for external connections to RS-232 connectors for serial input and output.

The **PIO** chip uses the same scheme of address bus connection and enabling line from the address decoder. It has a bidirectional connection with the data bus and also with the control bus, so that it also can send signals to the MPU to interrupt it and run an input routine if required. This chip also has the parallel input/output (I/O) connections to the parallel port connector. Though the PIO is usually operated with a printer as an output only, it can be used by other units bidirectionally, making it possible to connect scanners and disk drive units through the parallel connector.

All of the computer actions are determined by the software, and a computer without software is no more useful than a record player without records. A small amount of software is permanently fixed into the ROM of the computer system, and this can be used to allow a few keyboard actions and, more importantly, to allow more software to be read into the memory from a magnetic disk, the **hard drive** (or from a large memory chip in place of a mechanical drive). This software is the **operating system**, which consists of a set of routines for using other programs, allowing all the effects that we take for granted. See Chapter 15 for more about operating systems. Operating systems and applications software, and the details of constructing computer systems, are outside the scope of this book, but if you are interested, take a look at my book *Build and Upgrade Your Own PC* (currently in its fourth edition). There are also many websites that will instruct you in the art of computer assembly.

■ Note

Microprocessors and computing methods are used in many applications that are not quite so obvious. The engine management system of a modern car is a microprocessor-based computer system that takes inputs such as engine speed, air temperature, throttle opening, and so on to control fuel injection and ignition timing. Automatic gearbox systems are similarly controlled and integrated with engine management. Even the humble washing machine or dishwasher is now certain to use a microprocessor system to replace the old methods based on electric clock mechanisms with cams and switches.

Summary

Calculators and computers use a microprocessor as the main programmable unit. Calculators generally use only a number keyboard and a number display, with a small memory. Computers use a much wider range of connections such as hard and floppy drives, compact disc (CD) units, sound boards, scanners, light pens, full-sized keyboards, and cathode-ray tube (CRT) monitors.

Peripherals

Modem

Definition

The word modem is derived from the words *mod*ulation and *dem*odulation, and is used for a device that allows computer data to be passed along telephone lines or by radio links. A telephone line was never intended for more than voice signals, so that it operates best with frequencies in the range of 300–400 Hz. By contrast, computer signals consist of successive 0 and 1 voltage levels which may be at a speed of many megahertz.

Clearly, it is impossible to transfer unaltered computer signals along an ordinary telephone cable. To achieve communication of data along telephone lines the data rate must be reduced, and some form of modulation must be used. This is the basis of the modulation action of the modem, and a block diagram of how the modem is used to link computers is illustrated in Figure 14.2.

Definition

Any sinewave carrier can be modulated in three different ways, by modulating its amplitude, its frequency, or its phase. This gives rise to the different methods that are used in modems, called amplitude shift keying, frequency shift keying, and phase shift keying. Combinations of keying methods are possible.

The **amplitude shift keying** (ASK) system alters the amplitude of the carrier according to the level of the data signal. At its simplest, the carrier can simply be switched on for logic 1 and off for logic 0 (this is also known as on–off keying or OOK). This type of modulation has fallen out of use because of its poor signal-to-noise (S/N) ratio and therefore its high bit error rates. The problem is that ASK signals and noise signals are very similar, so that a receiver cannot distinguish genuine signals from noise signals.

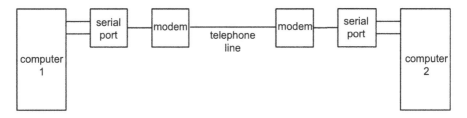

Figure 14.2:
Using modems to allow computers to communicate by telephone lines. This is the basis of electronic mail and the Internet

A **frequency shift keying** (FSK) system involves switching the carrier wave between two frequencies. In the 1970s, small computers used one form of this, known as Kansas City (or CUTS) modulation, for tape-recording digital signals, and this form is a good example of the method. For the CUTS standard bursts of eight cycles of 2400 Hz or four cycles of 1200 Hz are used to represent 1 or 0, respectively. FSK systems for modems have used other frequency values, often with a frequency ratio that is not exactly 2:1. FSK has a better bit error rate than ASK under the same conditions, but requires a wider transmission bandwidth.

Phase shift keying (PSK) is a method that is now the most commonly used for low-speed modems. This uses a single carrier frequency whose phase is altered by the data signal. For example, a logic 0 might cause no phase change, but a logic 1 will shift the carrier phase by 180°. Of the three methods, PSK has by far the best lowest bit error rate, and the bandwidth that it requires is least.

A more modern development of PSK is **multiphase PSK**, in which the carrier phase can be shifted to several different values. For example, if eight different values of phase can be used, then each digital signal represented by a phase change can consist of 3 bits (since a 3-bit signal can convey number values of 0 to 7, eight in all). If phase and amplitude modulation are combined, it is possible to carry 4 or more bits of digital data per unit carrier. Such systems are also used for the NICAM stereo sound system and for digital television (see later).

This allows a carrier of, say, 350 Hz to convey much more information than might be expected, and bit rates of 28,000 per second or more are normal. The effective bit rate can be increased by using data compression (eliminating redundant bits in signals) and by using faster rates along with error detecting and correcting methods. Older modems for small computers were designed to input information at 56,000 bits per second and output at a lower rate (because modems are used in small computers mainly to read in data, with very little being passed outwards). Modems are now used to link computers (using the Internet) that can be located anywhere in the world, and pass information over any distance.

The traditional type of modem for many years was the dial-up type, but around 2000 broadband began to take root in the USA (the UK date was around 2002). Broadband overcomes two serious limitations of the dial-up system, one of which is that it ties up the telephone line so that when the modem is in use the telephone cannot be used. The other disadvantage is the limited speed of the dial-up modem. Using broadband, speeds of up to 8 megabits per second are possible if you live reasonably close to a telephone exchange, and much higher speeds are possible if your telephone is supplied by fiberoptic cable.

Dial-up principles

In practice, a modem is connected to the serial port of the computer, or plugged in to the bus lines. Software is used to set up the modem and the port for the required data rate, the telephone number to use, and so on. When a communication is to be sent, the software will make the modem open the line and dial the number, and when the remote computer answers, the two confirm connection and the data is sent. At the other end, the modem has responded to the ringing tone and has opened the line. The signals from the transmitting modem are analyzed, the receiving modem automatically sets itself to use the same system. It then acknowledges contact, and receives the signals, which are usually stored as a file on the disk. The modem can be located inside the computer or externally.

■ Note

If you use the dial-up system with just one telephone line to your house, you cannot use the telephone at the same time as you use the Internet. At one time, some users would install two telephone lines, keeping one for voice calls and the other for the Internet. With broadband, you can use both together.

Summary

The dial-up modem (modulator—demodulator) is used to connect computers over telephone lines. A low-frequency carrier is modulated by the digital signals from a low-speed serial port. By using suitable modulation systems, rates of 9600 bits per second and higher can be achieved with a very low error rate. Modern software allows the modem to be controlled very easily, and enables actions such as automatic dialing and answering. The use of broadband has superseded dial-up now for most users.

Broadband principles

Broadband works by using separate frequency ranges over the telephone lines. The lowest frequencies are used exclusively for telephone conversations. The higher frequencies are reserved for broadband signals, and at each telephone socket you need to use a microfilter that separates the frequencies. The microfilter plugs into the phone socket, and your telephone plugs into the telephone socket on the microfilter. Your computer is connected (by way of a modem or, more usually, a modem—router) to the other outlet on the microfilter. The system that is used for domestic broadband is called ADSL (asymmetric digital subscriber line), and the asymmetric part refers to the rates: fast for downloading data to your computer, but slow for uploading from your computer to a website (since you need downloading much more than uploading). The arrangement of frequency bands is illustrated in Figure 14.3.

■ Note

Broadband can also be used over other metal lines, such as power lines.

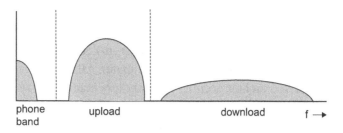

Figure 14.3:
The arrangement of frequency bands for broadband

The predominant system for modulation in the UK is called **DMT** (**discrete multitone**). As the name suggests, the available bandwidth for ADSL is used by a set of sub-carriers that are spaced 4.3125 kHz apart. Each of these sub-carriers will be modulated by a portion of the digital data, and sub-carriers that are not required will be suppressed so as to save total bandwidth. This approach is similar in principle to the methods used for digital radio, and the usual digital processing methods are used to ensure high efficiency and signal integrity.

Given then that you can get the modulated broadband signals from the microfilter, what then? You could use a broadband modem, but most users nowadays go one step further and install a **modem−router**. As the name suggests, the router connects to the broadband socket of the microfilter and passes demodulated data signals to and from one or more computers. The advantage of this is that you can connect more than one computer to the broadband (and to each other), so setting up a home network. In addition, if the router is of the wireless type you have the choice of connecting by way of a WiFi signal as well as by cables.

The most significant advantage, however, is in security. Modems are not truly secure devices, but routers are much better and if you enable the firewall of the router and use the highest level of security for any WiFi connection then you can be reasonably relaxed about security (though Windows users should ensure that they have an effective virus detector such as AVG Free).

Though ADSL through existing telephone lines is the most favored system in the UK at the time of writing, delivering speeds up to 8 Mb/s in areas close to a telephone exchange, it is not the only method of obtaining broadband. In areas that are serviced by cable television (using fiberoptic cable with wired connection to each household) higher speeds and more reliable connection can be achieved. At the time of writing there is a proposal to upgrade all existing telephone lines, starting with a few selected towns, to fiberoptic cables that are laid in the same conduits as the existing copper lines. This should bring much higher speeds (50 Mb/s or more) for these areas, and the hope is that this will be extended over much of the UK.

Another broadband system uses the mobile phone network. The cost may based on the amount of data downloaded each month (ranging from 5 GB upwards) rather than on speed, or a combination of speed (typically 7 Mb/s) and download limit (0.5−5.0 GB typically) can be quoted. The number of providers and options makes it very difficult to decide what is a reasonable deal for any given user. The technology is simple: a universal serial bus (USB) fitting adapter (a **dongle**) contains the mobile phone SIMM chip (single in-line memory module; the data chip for the mobile phone) and is plugged in to your computer. All the software supply and setting up are then carried out for you and the security risks are less than for a fixed line system. This type of connection is intended primarily for those who must use the Internet on the move (despite the signal fluctuations from place to place) and the cost makes it less attractive to home users.

The last option is broadband by satellite, using the incoming satellite signals to carry the downloaded data and a fixed telephone line to upload. Prices are comparable to ADSL, with the advantage that the signal can be better in areas that are distant from a telephone exchange. For business users for whom cost is less of a factor, a transmitting satellite dish can be used to provide faster uploading.

Monitor or Visual Display Unit

Definition

The monitor or VDU is the system for displaying an image. Early microcomputers made use of a television receiver, but the need for higher resolution has led to the development of specialized displays which are nowadays always color units.

The conventional older type of monitor uses a CRT or liquid crystal display (LCD) panel to produce a picture that consists of a set of colored points. The **resolution** (meaning how much detail can be seen in a picture) can be measured in terms of the number of dots per screen width and depth. Note that a printer achieves much higher resolution than any monitor, measured in terms of dots per inch rather than dots per page width. The equivalent dots per inch for a screen is around 75; the accepted minimum for printers is 300 and many printers achieve 600 dots per inch or more in monochrome (a lower figure for color).

The type of picture that we can show on a monitor depends greatly on how much control we can achieve over the brightness and color of each individual dot on the screen, and this is where the **graphics card** becomes important, because when the personal computer (PC) first appeared it could not display any graphics, only text. This required only minimal control, light or dark for the letters and symbols of the alphabet and digits.

The simplest way of controlling the dot brightness on a monitor is simply to turn the dot on or off, so that **on** means bright and **off** means dark. This scheme was used for monochrome monitors before color pictures became the universal standard, and it was later extended to color displays by using on—off control for each of the primary colors of light, red, green, and blue. Signals that use the three color primaries in this way can produce red, green, and blue using the dots for primary light colors and can also show the mixture of red and green, which is yellow, the mixture of red and blue, which is magenta, and the mixture of blue and green, which is called cyan. The mixture of all three colors is white. Using a simple on–off scheme like this can therefore produce eight colors (counting black and white as colors).

The simple on–off method can be improved by using a fourth on–off signal, called brightness or **luminance**, the effect of which is to make the colors brighter when this signal is switched on, so that you have black and gray, red and bright red, and so on, a range of 16 colors including black. This 16-color system was so common in the past that a lot of software still features 16 colors despite using a graphics system that allows a much greater range of colors. A monitor of this type is a digital monitor because its inputs are digital.

Analog monitor

The alternative to these on–off video signals is to use the **analog** type of signal that is used for television monitors, in which each signal for a color can take any of a range of sizes or amplitudes, not just on or off. This method allows you to create any color, natural or unnatural, by controlling the brightness of each of the primary colors individually. A monitor like this is much closer to television monitors in design, and some (but certainly not all) monitors of this type can be used to display pictures from sources such as television camcorders and video recorders. Monitors for a modern PC should allow the use of resolutions up to around 1600×1200. The color capabilities depend on the number of bits used to code each pixel, and the relationship is that the number of colors that can be displayed is 2^N, where N is the number of bits per pixel. Twenty-four-bit color is a common requirement for applications such as digital photography, and this corresponds to 16,777,216 colors, usually referred to as 16 million colors. Thirty-bit and 32-bit color systems are also used for critical applications.

■ Note

Some older computer monitors are referred to as **digital**, but this does not mean that the signals are digital. In this context, digital means that there are no potentiometers used for adjustment of such items as brightness or contrast, only push-button switches (one for increase, one for decrease). This use of the word 'digital' has now died out because all monitors now use selection by push-button or menu selection.

■

Summary

Monitors originally used a digital form of signal, displaying a monochrome picture for text use. Later, a 16-color system was developed by using digital switching of each of the primary colors, plus a brightness signal. All of these display types are now obsolete, but modern monitors can use a digital connection that produces a better picture quality than the analog type of connection. Modern monitors, mostly LCD types unless for very large displays, are capable of high resolution and a full range of color, 24 million colors or more.

Graphics Card

Definition

The signals from a computer cannot be used for a display directly, and a graphics card, plugging into a slot inside the computer, is used to perform this conversion and to produce signals of the correct type and voltage range for a monitor.

■ **Note**

Some modern computers have **motherboards** that include the circuits of a graphics card, and many also include a sound card.

■

The original type of PC display was concerned only with text, because the concept of a machine for business use at that time was that text, along with a limited range of additional symbols, was all that was needed for serious use as distinct from games. The original video card was referred to as the **monochrome display adapter**, a good summary of its intentions and uses, and usually abbreviated to MDA.

MDA produced an excellent display of text, with each character built up on a 14 × 9 grid. The text was of 80 characters per line and 25 lines per screen, and at a time when many small computers displayed only 40 characters per line using a 9 × 8 grid, this made text on the IBM monitor look notably crisp and clear. Typical monitors used a 18.4 kHz horizontal scan rate and 1000 lines resolution at the center of the screen, and they still look good for a text display.

Graphics pictures could not be displayed using the MDA card, however, so that graphs could not be produced from spreadsheets (on screen at least), and applications such as desk-top publishing (DTP), painting, digital photographic editing, or computer-aided design (CAD) were totally out of the question. As the need for graphics in color grew, various solutions in the form of different graphics cards appeared. The 16-color color graphics adapter (CGA) was used in the mid-1980s, but it was soon replaced, first by the extended graphics adapter (EGA) and then by the video graphics adapter (VGA) cards.

■ **Note**

The VGA standard permits full compatibility with earlier types of display, and it adds displays of 640 × 480 16-color graphics and 720 × 400-color graphics, using a 9 × 16 grid for characters with color. The VGA type of card used nowadays is classed as SVGA (S for super), and it permits much higher resolution (such as 1280 × 1024) and the use of a large number of colors (16 million or more). You should not consider using an older type of graphics card on a modern PC, because some modern programs would not run on an older type of card, and others would provide very disappointing graphics. One particularly useful action of a graphics card is to scale large images to the screen resolution, so that a camera image of around 4000 × 3000 can be displayed on the whole screen of 1280 × 1024.

■

Definition

All modern graphics cards contain a specialized processor, the **graphics processor**, along with chips to support the fast actions that must be performed so that a set of digital data can be transformed into the signals for a color image. A substantial amount of memory is also needed.

The more memory your graphics card can use, the more easily it can work with high-resolution and high-color images, and some graphics cards come with 128 Mbyte or more of their own memory. Others make do with less, and some will grab memory from the computer's RAM, which is not an ideal situation unless you are using several gigabytes of RAM.

Another point to consider is that it takes time to transfer the large numbers of bytes that images require, so that the speed of a graphics board is important. To achieve faster rates of transfer of bytes, later graphics cards used slots connected so that they could operate at higher speeds. Originally, a slot type termed PCI, operating at 33 MHz, was often used, but faster types such as AGP (66 MHz) replaced it. The most recent type of slot at the time of writing is the PCIe, with the 'e' meaning express. This is a very fast serial type of connection with the computer, introduced in 2004, and it is likely to be found on modern computers, albeit in a variety of versions ranging (in order of increasing speed) from PCIe-1 to PCIe 16/2.

The graphics card must also be connected so that it can send signals to the monitor, and the conventional VGA type of connection was designed for CRT analog monitors, though its use continued on LCD monitors. Modern LCD monitors use the digital video interface (DVI) type of connector, though the VGA connection is often provided as well. The two are impossible to mix up because the VGA type uses a 15-pin connector with round pins and the DVI type uses a different pattern of flat pins. Just to add to the confusion, however, these cables come in the legendary 57 varieties, and it is possible to buy a DVI cable and connector that will connect with both analog and digital monitor inputs. As always, you have to specify what graphics card you want to connect to what monitor.

Summary

A graphics card is the interface between the computer and the monitor. It contains a processor of its own (a graphics processor), along with support chips and memory, often as much as 4 Gbyte (though only the fastest and most expensive graphics cards can make use of such large memory sizes). You have to specify the correct type of connection to your computer, and the correct type of connection from the graphics card to the monitor.

Definition

A **driver** is a program that is used to control the signal output from the computer to a device such as a monitor, printer, mouse, keyboard, or scanner. The PC operating system, typically Windows, provides a set of drivers for the most commonly used graphics cards, and also for several that you are not likely to encounter in the UK.

As always, nothing on a computer works unless there is a program for the action, so that graphics boards need such a program, called a **graphics driver**. Each design of graphics board will need to have a driver written for it, and these drivers can often be a source of problems if they interfere with other computer actions.

If you use an operating system other than Windows (such as Linux or Mac) you should make sure that a driver is obtainable before you buy any equipment that needs a driver.

Sound Card

Definition

A sound card (or sound board) is used as a way of allowing the computer to work with digital sound signals, such as the output from a music CD. The sound board contains A/D and D/A converters, so that it allows a microphone or other source of sound signals to be connected as inputs, or headphones or loudspeakers to be used as outputs. The use of the microphone also permits direct voice input software to be run. A sound board is an essential addition if multimedia CD-ROMs are to be used.

Early sound boards or cards used 8-bit single-channel sound coding, and this is too crude for music of reasonable quality, though adequate for speech or sound effects. Later boards all use 16-bit sound, allowing for music to CD standards. In addition, the later boards are two-channel for stereo sound. Such boards can be very elaborate, incorporating specialized chips for manipulating digital sound. Many card types come with software for editing sound files.

The facilities you can expect from any sound board include a CD-ROM drive interface, loudspeaker and amplifier outputs, inputs for microphone (high sensitivity) and for other (stereo) sources (line input, lower sensitivity), a volume control for use when passive loudspeakers are used, a musical instrument digital interface (MIDI; which in some designs can also be used for a joystick), and stereo headphone connections. As usual, the sound card needs a software driver.

■ Note

Loudspeakers for connecting to a sound card are almost always of the type called **active loudspeakers,** which means that they do not rely on the power of signals from the computer. They come with a mains adapter that provides power independent of the computer.

One of the two active loudspeakers contains an amplifier, and has to be connected to the house mains supply. It also sends the amplified signal to the other loudspeaker. The cable connecting the computer to the loudspeakers is usually fitted with miniature jack plugs at either end.

Summary

The sound card is added to the PC, or it can be incorporated on the main board (motherboard). It provides connectors for input and output of analog or digital sound, along with analog to digital (A/D) and digital to analog (D/A) converters. The loudspeaker output of the sound card is normally used for a pair of active loudspeakers.

Printer

Definition

A printer is an important computer peripheral, a device that is external to the computer and connected through a data cable. The printer allows text or graphics to be printed on paper.

Many types of printer have been marketed for computers, but the dominant varieties are inkjet types and laser printers. The impact dot-matrix printer that was once the only type available is rarely seen nowadays, and the daisywheel (like an electric typewriter) is obsolete. All printers require a considerable amount of processing ability to convert the signals of codes from the computer into marks on paper, so that a printer will always contain a microprocessor and some memory. In addition, power transistors are used for controlling electric motors and other aspects of printing and paper handling.

At one time all printers used the old standard Centronics connector, so that you could use any printer with any computer, provided that software drivers were obtainable. At the time of writing, USB connectors have replaced the old (and bulky) Centronics type. These connectors are illustrated in Figure 14.4. Some printers also use network or radio connections so that they can be operated from more than one computer or at a distance.

Paper feeding is tractor, sheet-feed, or manual. Tractor feed, now used mainly for office equipment, uses paper that is perforated with holes of about 4 mm diameter on the margins so that it can be moved by a pair of wheels that have teeth which engage into the holes of the paper. The margins are lightly perforated for easy removal, and the continuous paper can be separated into separate sheets (burst) by tearing along the divisions. Sheet-feed uses a hopper to carry the paper, usually 50 sheets or more, and each sheet is fed on demand into the printer. Manual feed requires you to insert each sheet by hand and prompt the printer to work on it. By convention, tractor feed paper is used mainly in US sizes, but sheet-feed normally uses A4 in the UK. Sheet-feed is almost universal on printers intended for home use, but several printers cater for inserting single sheets or cards even when the main feed is from a hopper.

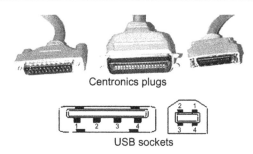

Centronics plugs

USB sockets

Figure 14.4:
Standard Centronics parallel printer connector along with the later USB connection (not to scale) that has replaced Centronics

Inkjet printer

Definition

The inkjet printer uses a matrix principle, with a line of tiny jets used to squirt a fast-drying ink at the paper. A line of dots is placed on the paper and the inkhead then moves a fraction of a millimeter across the paper to deposit the next line of dots. This process is repeated until the entire shape of a letter is printed. The two main systems currently are the bubblejet and the piezoelectric jet. Both need power transistors to actuate the printing of each dot.

The bubblejet system, devised by Canon, uses a tiny heating element in each jet tube, and an electric current passing through this element will momentarily vaporize the ink, creating a bubble that expels a drop of ink through the jet. The piezoelectric system, devised by Epson for its Stylus™ models, uses for part of the tube a piezoelectric crystal material which contracts when an electrical voltage is applied, so expelling a drop of ink from the jet. This latter system has been developed to allow for high-resolution printing of 720 dots per inch or more, and bubblejet models of high resolution are also available.

Both types of inkjet can print in color (using colored inks in addition to black), with resolution of 600 dots per inch or more, and at speeds of, typically, 12 pages per minute of text. Color printing is always much slower than monochrome. Inkjet printers are cheap, but the manufacturers recover their investment by charging very high prices for the ink cartridges. Combination inkjet printer/scanner machines (often referred to as all-in-one) can be used as copiers.

Laser printer

Definition

The laser printer has become the standard office printer on the grounds of high-speed printing and quiet operation. The principle is totally different from that of inkjet printers, and is much closer to the photocopier principle, based on xerography.

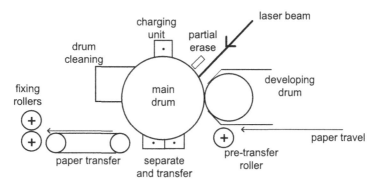

Figure 14.5:
The outline of a monochrome laser printer

The heart of a laser printer is a drum which is wider than the width of the paper the machine will use and made from light-sensitive material. This drum is an insulator, so that it can be electrically charged, but the electric charge will leak away in places where the drum has been struck by light. The principle of the laser printer (and the photocopier) is to charge the drum completely and then make the drum conductive in selected parts by being struck by a laser beam. The beam is switched on or off and scanned across the drum as the drum rotates, all controlled by the pattern of signals held in the memory of the printer, and enough memory must be present to store information for a complete page. A schematic for a typical laser printer is shown in Figure 14.5.

This system requires about 0.5 Mbyte of memory as a minimum for text work, and 2 Mbyte or more if elaborate high-resolution graphics patterns have to be printed. Color laser printers can use typically 64 MB maximum. Once the scanning process is complete, the drum will contain on its surface an electrical voltage 'image' corresponding exactly to the pattern that exists in the memory, which in turn corresponds to the pattern of black dots that will make up the image. Finely powdered resin, the **toner**, will now be coated over the drum and will stick to it only where the electric charge is large, i.e. at each black dot of the original image.

■ Note

Color laser printers are obtainable, and are no longer costly. The cartridges are very expensive, but they have a very long life compared to the inkjet cartridges (typically, my black cartridge is replaced once a year and the color ones every three years) so that maintenance costs compare well with inkjets. However, it may be cheaper to replace the printer rather than buy a complete set of replacement cartridges, because the cartridges are very large and elaborate compared with inkjet types. Another point is that you can operate a laser printer for years without problems, whereas inkjet types tend to suffer from blocked jets and, once again, it is cheaper to buy a new inkjet printer than to replace the jet heads. Hewlett-Packard printers have combined ink-tanks and jetheads, so that the heads are replaced as well as the ink. This is more costly than replacing an ink-tank, but it avoids the problem of having to replace a set of blocked heads.

The coating process is done by using another roller, the **developing cylinder**, which is in contact with the **toner** powder, a form of dry ink. The toner is a light dry powder which is a non-conductor and also magnetic (some machines use a separate magnetic developer powder), and the developing cylinder is magnetized to ensure that it will be coated with toner as it revolves in contact with the toner from the cartridge. A scraper blade ensures that the coating is even. As the developing cylinder rolls close to the main drum, toner will be attracted across where the drum is electrically charged, relying on the electrical attraction being stronger than the magnetic attraction. Note that two forms of attraction, electrostatic and magnetic, are being used here.

Rolling a sheet of paper over the drum will now pass the toner to the paper, using a corona discharge to attract the toner particles to the paper by placing a positive charge on to the paper. After the toner has been transferred, the charge on the paper has to be neutralized to prevent the paper from remaining wrapped round the drum, and this is done by the static-eliminator blade. This leaves the toner only very loosely adhering to the paper, and it needs to be fixed permanently into place by passing the paper between hot rollers which melt the toner into the paper, giving the glossy appearance that is the mark of a good laser printer. The drum is then cleared of any residual toner by a sweeping blade, recharged, and made ready for the next page.

The main consumables of this process are the toner and the drum. The toner for laser printers is contained in a replaceable cartridge, avoiding the need to decant this very fine powder from one container to another. The resin is comparatively harmless, but all fine powders are a risk to the lungs and also carry a risk of explosion. Drum replacement will, on average, be needed after each 80,000 copies, and less major maintenance after every 20,000 copies. Paper costs can be low because any paper that is suitable for copier use can be used; there is little advantage in using expensive paper, and some heavy-grade paper may cause problems of sticking in the rollers.

Some models use a separate developer powder (a set of negatively charged and magnetic beads) in addition to toner, and the developer will have to be replenished at some time when the toner is also exhausted. The Hewlett-Packard LaserJet machines use a cartridge which contains both the photoconductive drum and the toner in one package, avoiding the need for separate renewal. The life is typically quoted at about 3500 sides at the average print density of word-processed text, but this figure will be drastically reduced if you print a lot of dense graphics and fonts. Long-life cartridges are available.

■ Note

A monochrome printer of any type cannot reproduce true continuous-tone photo-graphs, because each dot that it prints is black. The effect can be simulated by mixing black dots and white spaces, a process known as half-toning, but this leads to a coarse appearance on a 300 dot per inch printer and is really satisfactory only on a typesetting machine which works at 1200–2400 dots per inch. Note also that the heating element (fuser) used in a laser printer will cause problems of bubbling with glossy paper (such as is used for printing photos on inkjets). Color laser printers will give a glossy finish on plain papers and the very glossy type should be avoided.

■ Note

All printers can work with ordinary office paper provided it is not too thick. In some printers, particularly color laser printers, the path of the paper through the printer is so tortuous that only relatively thin paper can be used, so that printing items such as business cards cannot be done. A few printers feature a straight paper path, and this is very much better if you are likely to want to use thicker material.

Summary

Printers for small computers are either inkjet or laser types. Inkjet printers are very versatile, and are well suited to digital photographic work as well as for normal text printing. Laser printers are more appropriate when a large volume of printing work is required. Whatever type of printer you use, a suitable software driver must be installed in the computer.

Scanner

Definition

A scanner is a device that reads documents by using an image sensor strip, and stores a digitized image as a file. This can be used as a form of copier, for editing graphics, for fax transmissions, or for reading text into a word-processor without retyping. Scanners can be handheld, flatbed or roller-feed types, though flatbed types are by now the most common and the others are seldom offered for sale. More recently, combined scanner/printer units have become popular.

■ Note

Scanners are connected to the computer nowadays by way of the USB socket, which has replaced a variety of older connectors. Whatever method is used, driver software will have to be installed before the scanner can be used. Some scanners (such as the Canon LIDE series) need no separate mains cable because they can use the 5 V supply that comes from the computer through the USB cable.

The item to be scanned can be any mixture of text and images on paper, and modern flatbed scanners all work in color, usually 24-bit. A cold-cathode light source in the shape of a tube is moved over the document, and the reflected light is sensed by a set of photosensitive dot-sized cells, each of which contributes one pixel of the image. The speed of movement is controlled by a servomotor, and resolution can be increased by reducing the speed of movement and by making more than one pass over the document with a small displacement from the original position. Low-power scanners use a light bar of tiny light-emitting diodes (LEDs) that require much lower power than the old-style light tube.

For scanning images, the scanner can be controlled from within any modern graphics program, such as **Paint Shop Pro** or **GIMP**, and the resulting digitized image can be stored, transmitted through the Internet, or printed as required. More specialized software is needed if the words on a page are to be converted into the form of a word-processor file, because the scanned file is a bitmap, whereas a word-processor uses one byte for each character.

Definition

OCR means **optical character recognition**, and refers to software that will recognize the graphics image of a character and convert this into the ASCII (American Standard Code for Information Interchange) code for that character. In other words, OCR allows the scanner to be used as a text reader, providing an input to a word-processor.

OCR software is usually bundled with a scanner, and provides for a very high standard of text recognition, so that only a spell check is usually needed on the text after recognition. Problems can be experienced with very small character sizes, italic text, or tables, or with a document received by fax. You can usually opt to vary the resolution for OCR from 200 dots per inch, suitable for large type sizes, to 400 dots per inch, suitable for smaller text.

A flatbed scanner holds the page steady and scans it, using a system very like the scanner of a photocopier. The quality is excellent, and modern models allow for color scanning at high resolution and 24-bit color. Automatic page feed can be used on some types, which is particularly useful when OCR is being applied to a long document. The cheapest scanners are useful for domestic purposes, but for professional purposes much more expensive and elaborate machines are available.

Summary

A scanner is a valuable peripheral that allows you to digitize documents that consist of text and/or images. Scanned text can use OCR software to convert it to ASCII text files for a word-processor. Images can be scanned at high resolution and in 24-bit color using modern flatbed scanners.

CHAPTER 15

Computer Construction and Software

Computer Types

Definition

Computers, as applied to domestic use, exist in desktop, laptop, notepad/notebook, and tablet sizes and versions. The desktop types use separate computer, monitor, mouse and keyboard units connected by cables; the laptop, notepad, and tablet types are integrated into one unit, though the sensitive pad used in place of a mouse on laptops and notepads is so generally disliked that provision is usually made for attaching a mouse through the universal serial bus (USB) port. Network connection can be by RJ45 cables or by radio (WiFi), and laptops usually feature both connections. The smallest notepad or netbook computers use radio connection only, and machines that have no provision for networking can generally be converted by a USB-fitting adapter. The tablet type is, at the time of writing, new, and was introduced by Apple, though many other manufacturers are following suit. The tablet in its usual form has no keyboard, and uses a stylus and a touch-sensitive screen. A few models allow the use of a plug-in keyboard.

Constructionally, the main divisions for computers for domestic use are PC, Mac, or ARM. The PC type of construction normally uses processors by Intel or AMD (though a few other types are sometimes specified) and follows the general pattern set by the original IBM PC in 1982 and considerably enhanced since then. Mac (Apple) machines are now constructed using Intel hardware (but in the past have used Motorola and PowerPC chips). Laptops are either PC or Mac in construction, as also are the netbook machines such as are manufactured by ASUS and Acer. The much more recent micronetbook machines (usually in the sub-£100 price range) use the ARM type of processor, and this type of processor is also the processor of choice for tablet computers. Though anyone can construct a PC (or the ARM type), the Mac is constructed and supplied only by Apple, which allows the price to be maintained.

Laptops and Netbooks

The difference between a laptop and a netbook/netpad is a matter of size and facilities. A modern laptop is a fully fledged computer in a compact form, with the speed, memory size, and drive capacity that you would expect from a desktop machine. It can be operated from mains power or from a built-in battery (for a few hours). The netbook is intended as an

Electronics Simplified. DOI: 10.1016/B978-0-08-097063-9.10015-9

accessory to a desktop or laptop, often with a faster boot-up time, but with limited memory, smaller screen, battery life, and hard drive size; quite often the hard drive may be replaced by a solid-state drive (SSD) so that the computer is much quieter than a laptop. Netbook machines also usually dispense with an optical drive, so that new software has to be loaded from the Internet or from a flash drive (pen drive).

The netbook market has for some time been served by processors from Intel (Intel Atom) and from AMD (AMD Neo, though a new processor is rumored at the time of writing). Both of these are low power (in the wattage sense, so as to prolong battery life) but run at quite high clock rates, typically 1.6 GHz. Both are being used in netbook applications, though AMD have insisted that the Neo is not aimed at this market, despite a version of the ASUS Eee appearing with this processor. New processors for this important market can be expected to appear over the next few years.

■ Note

The netbooks from manufacturers such as ASUS and Acer are in several forms, some using a version of Windows as the operating system (OS), others using Linpus Linux Lite. For the smaller machines with an SSD and small memory, the Linux system is very much superior, allowing fast booting and having a good range of applications built in. You can also download and install other software, and even install another OS such as Ubuntu Notebook Remix (UNR), Ubuntu Mini Remix, or Windows 7.

The ARM type of machine, typified by the first netbooks selling for under £100, uses a processor that can be identified as ARM or VIA. The ARM chip is a 32-bit processor that has been developed from the processor used for the Acorn, a machine much loved in its time. The chip is of the type known as RISC, meaning reduced instruction-set computer to reflect the point that it used a small number of simple instructions as distinct from a large set of complex instructions as found in the Intel and AMD processors.

Most of the mobile phone manufacturers worldwide use a form of ARM chip, and they are also used in calculators, game consoles, hard drives, and routers, to name just a few of the applications for this versatile chip which can be customized for any specific application. The ARM processor technology has also been licensed to other manufacturers who use different names. Later versions of the ARM processor are now the processor of choice for the new tablet computers.

■ Note

One snag of some of the miniature ARM machines is that they have all of their software in read-only memory (ROM) so that applications cannot be added or removed.

Taking a typical netbook/netpad use of ARM, the processor works at the low clock speed (by modern standards) of around 400 MHz, and though the processing speed is adequate for requirements such as document writing and display, photo display, video display, Internet browsing, and spreadsheet use, it cannot cope with such activities as video editing and any use of downloaded software (such as applications like YouTube or Facebook). It is suitable for anyone who has a desktop or fully featured laptop as the main computer and who will use the netbook as a way of working on office documents away from home. The OS is usually Windows CE-6, of which more later. Memory can be added by way of a camera-style secure digital (SD) card, and the USB sockets permit the use of attachments such as memory sticks, optical drive or hard drive and other USB connected items (but **not** a USB WiFi). For anyone who has been accustomed to using Windows (XP onwards) the CE version can appear very strange and clumsy.

■ **Note**

Some of the early sub-£100 netbooks have arrived with serious flaws that make them almost unusable except for writing documents (but not printing, because there are no printer drivers) and displaying photographs. By the time this book is in print the market should have sorted out the winners from the losers, but if you try out one of these machines be sure to test it thoroughly before the guarantee (if there is one) runs out. Tablet machines can also present problems, particularly if you want to create documents, since the absence of a keyboard means that the machine must recognize handwriting or use a touch-screen image of a keyboard.

■

Summary

Computers for home use fall into four categories of desktop, laptop, notepad/notebook, and tablet. Though laptops can be as capable as desktops, their price for the same specification will inevitably be higher. Notepad machines are regarded as auxiliaries to desktop or laptop machines, though some can be almost as capable for office work that does not require optical drives, high speeds, or large memory. The tablet type of machine is aimed at those who need computing facilities on the move. If needed, however, an optical drive can be added using an external casing with a USB connection, and memory can be added by way of SD cards.

Only the PC type of desktop machine offers the chance to build or repair your own computer easily, because all of the parts that are needed can be bought on the open market. The main parts are the case, the motherboard, and the processor, and you will find case/motherboard packages on offer. It is a fine piece of judgment, because you need to buy items that are compatible and which will not become out of date too quickly, yet sufficiently established to be sure that any problems have been sorted out.

■ **Note**

One important point is that unless you need exceptional speed there is no point in paying out for the latest hardware. Many users run applications (word-processing, still-picture editing, video display, Internet) that could be tackled by any computer ten or more years ago.

■

If you go down the road of buying a combined case/motherboard package you will not have to cope with the most difficult choices and construction actions involved in fitting a motherboard to a case. You need only install the processor and, again, choice is important because the processor must be compatible with the motherboard, and the number of pins on a processor increases for each new generation of processors.

The remaining items are memory and disk drives, because the motherboard will incorporate such items as USB and other ports, and probably sound card and video graphics card as well. Memory is too important to buy from unnamed sources because it must be exactly matched to the rest of the hardware, and the memory selection sites for **Crucial** and **Kingston** can be recommended with confidence. The optical drive (usually a DVD rewriter) and hard drive(s) can be bought from any supplier as long as you keep with well-known makes.

If you are interested in constructing a PC to your own specification, see the Newnes book *Build and Upgrade Your Own PC*, currently in its fourth edition.

Operating Systems

Definition

The **operating system** (**OS**) is the software that determines the capabilities of the computer hardware by providing a set of programs that control the hardware.

Each variety of the early microcomputers had its own systems of software for controlling the actions, such as keyboard, mouse, and monitor functions, and these systems were built into the hardware in the form of a ROM. When the dust settled, the advantage of using a single standardized OS became apparent, and the domination of the office market by the PC type of machine meant that a standard OS would be designed for that machine.

As OSs for computers developed, it became unreasonable to fix them in ROM, and the normal procedure became to have all software on a hard drive. An OS on a hard drive allows for changes, whether major (as a **new release**) or minor (as in **updates**), and it also permits the user to settle for an OS of his or her choice as opposed to one favored by the manufacturer. The PC has always allowed this choice, and it has more recently appeared on the Mac, but the miniature netbook or notepad machines, along with tablet types, usually have an OS (such as Windows CE) that is fixed in ROM.

■ Note

Though the OS for a desktop or laptop computer is read from a hard drive, there are countless devices that need an OS of some kind that is fixed in ROM. This is an **embedded** OS, and these are so commonplace at the time of writing that (like a rat) you are probably less than six feet from one right now. A typical example of a device using an embedded OS is the 3G cellphone, but this is just one example of an embedded OS that is obvious because it is attached to a keyboard and screen, and you tend not to see where others are embedded in domestic machinery (washing machines, automobiles, security systems, etc.).

■

The most familiar type of OS is the hard-drive (or SSD) type used on computers, but there is another type called RTOS, meaning real-time operating system. When you use a computer you often type on the keyboard and then print the resulting document. If this were happening in real time, the printer would be printing while you were typing (remember the old-style typewriters?). RTOS is used for devices such as cellphones, domestic appliance controllers, programmable thermostats, industrial robots, and other devices where the output has to appear at virtually the same time as the input.

The PC Machine

The original PC machine of the early 1980s was supplied with an OS called Microsoft MS-DOS (DOS meaning disk operating system). This was (and still is, because it still exists as part of Windows) a **command-line instruction** (**CLI**) system. To carry out an action, a command was typed, together with any required data, and the C/R key was pressed to implement the action. This OS followed the pattern established for larger (mainframe and minicomputer) machines, many of which used manufacturers' own OS.

From 1969 onwards, however, an OS called **Unix** had become the first choice for many types of large computers, though at the time it was much too large to use in microcomputers. Unix became the OS of choice when it permitted use by several terminals (**multiuser**) and the ability to run more than one application at a time (**multitasking**). The importance of Unix is that it has heavily influenced some types of OS for modern machines (see later).

The CLI is still regarded highly by computer professionals for its speed and the range of control that it permits over the machine. Research at Stanford University, later extended by Xerox Laboratories, developed an alternative, a scheme using windows, icons, and a mouse, that we are so familiar with today. This was implemented by Xerox in 1981, and was adopted for an Apple machine (Lisa) in 1983 and by machines from Commodore and Atari later. This type of OS is known now as a **graphics user interface** (**GUI**), rather than the Windows icon mouse processing (WIMP) title that was preferred by those who thought that any departure from CLI was a retrograde step. All OSs for small computers now permit use either as CLI or as GUI.

■ **Note**

The best known GUI systems for domestic use are Microsoft Windows, Apple Mac OS, and Linux (a modern version of Unix). All of these also allow for a **terminal** or **command prompt** option that permits CLI commands to be input.

■

When you buy a new computer you usually have little or no say about the OS. A PC will normally come with the most recent version of Microsoft Windows and a Mac with the most recent version of Mac OS. A few assemblers of computers will install the OS you want for a PC, which gives you the choice of Windows or Linux. If you are assembling your own computer or updating an old one you also have this choice, and it is worth pointing out that older machines that have difficulty running a modern version of Windows will often run very well on Linux (which comes in either 32-bit or 64-bit versions; use the 64-bit version only if you have a 64-bit processor). A further advantage is that several versions of Linux are distributed free, including a full suite of useful software.

■ **Note**

Another hangover from early CLI days is the use of **batch files** or **scripts**. These are text files that consist of CLI commands, and they are used wherever a set of actions has to be carried out repetitively or on a large set of inputs. The distinction is that batch files are peculiar to MS-DOS, whereas scripts can be carried out from within a GUI and are often written using a form of programming language such as Perl, Python, and even the programming language Java.

■

Windows

Windows has been the standard installed OS for PC machines since the first really usable version, Windows 95. Subsequent updates have been named in the sequence Windows 98, Windows NT, Windows 2000, Windows Me, Windows XP, Windows Vista and Windows 7. Windows 8 is expected in 2012. Of these, the sequence from Windows 95 to Windows XP has featured ever-increasing complexity and more demands on resources of memory size and processor speed, though Windows 7 comes in six versions of which the Starter is the least demanding but likely to be found only on notepad/netbook machines. The next up, Home Basic, is not available in the UK at the time of writing, and the appropriate version for home users in UK and Europe is Home Premium. Other versions are aimed at the professional user and are priced accordingly. If you are currently constructing or updating a PC, then a genuine XP Home Edition installation disk is probably the best option for price and capability.

What follows is a brief guide to installation. Note that OS versions now come on compact disc (CD) or digital versatile disc (DVD) because floppy drives are no longer fitted to computers. The snag with putting Windows XP on to a machine that you are building is that this version of Windows (like subsequent versions) will not work for more than 30 days until it has been *activated* by Microsoft. This is an antipiracy move, and the activation can be carried out over the Internet (automatically) or by telephone. Windows XP is set up to detect changes in your hardware, so that it will resist being copied to another computer, but a side-effect of this is that changes to your hard drive, memory, or other internal hardware may trigger a request for reactivation. This is not such a problem as you may think, and some users have reported that reactivation has not been needed even for a hard drive change. The reactivation procedure following a change of hardware (as distinct from a change of computer) seems to be reasonably flexible.

■ Note

If there is no existing version of Windows on your hard drive (and for a new drive this is quite certain), you will need the CD-ROM for the original equipment manufacturer (**OEM**) version of Windows, not the (cheaper) upgrade version.

■ Note

If you are about to install XP, make certain that you have made all the hardware changes to your computer that you intend to make for some time, because if you activate XP and then change the hardware you may need to activate XP again on the computer. This alone is a good reason for not using XP on a machine that you are likely to upgrade in future. If you do not activate XP immediately, you can use it for a month (with daily warnings about activation) and make hardware changes, then activate it. The examples shown here have occurred during an upgrade installation. Note that hardware changes mean internal fixed hardware such as graphics card, a network, hard drives, CD-ROM/DVD, and random-access memory (RAM), but not modems or devices connected through the USB ports.

The installation of Windows XP Home Edition is started by inserting the disc into your CD-ROM drive, and you may have to set the CMOS RAM (see later) to boot from the CD-ROM. You will be asked to choose one of three options, to install XP completely, to perform other tasks, or to check that your system is compatible. This last option is useful if you are installing into an old machine that has been upgraded; there should be no compatibility problems if you are using a fairly new machine (2005 or later) with components of known quality. If you run the compatibility check you may end up with a list of hardware that needs additional files to be installed. This list may not correspond to reality, because if old hardware has been removed it leaves files on the hard drive that the compatibility list senses as indicating the hardware being present. The additional tasks list contains the following options:

- Set up remote desktop connections
- Set up a network
- Transfer files and settings
- Browse the installation CD
- View the release notes
- Back

If you have files and settings on an old computer, you should use the 'Transfer files and settings' option to ensure that you have all the data and items such as e-mail and browser settings. You should also read over the release notes in case there is a reference to some problem that may arise.

All OS types have an initial display that guides you to other software, making it easy to find the applications that you want to use. Figure 15.1 shows this panel for Windows XP, obtained by clicking on the Start button that shows when you switch on.

■ Note

You should always keep a backup of all your files and settings at all times because these are the most difficult item to replace in the event of a crash that makes the computer unusable.

Once this has been attended to you can select the 'Back' option and Install XP. This is a completely automated process and by the time it finishes you will be running your new OS. The format of installing XP was carried over into Vista (not a universally liked OS) and Windows 7 with few significant changes.

Windows CE

Windows CE (or WinCE) is a cut-down and modified form of Windows intended for netbook computers, particularly for **embedded** OS (in ROM so that it cannot be altered by a virus).

Figure 15.1:
The Start button menu of Windows XP

Though the software looks familiar, the usage is not always straightforward, and tasks that seem everyday on desktop Windows, such as dragging a file from one part of memory to another, are complex or, in some cases, nearly impossible. WinCE can be implemented on any computer using a processor of the Intel x86 family, and also (in a different version) on ARM machines (making it the OS of choice for netbook machines).

WinCE is intended to run in a very small memory (by modern standards, remembering that my first computer used a whole 32 kB of RAM) of less than 1 MB. The netbook type of computer can operate with no form of disk drive, with the OS and all the available user RAM in solid-state memory. Such machines rely on devices such as SD memory cards and USB memory if more memory is required.

Apple Mac

The current version of Mac OS, at the time of writing, is OS X 10.6.6, and you can buy this version if you own an older machine. You cannot, in theory, install OS X on a PC, but as you might expect, someone has come up with a method. If you take this path, you can expect any problems you encounter to be ignored by both PC and Mac advisers. You can also run Windows on a Mac, but until very recently this was equally hazardous. You can run Linux on either type of hardware, but that is something we will look at later.

The old generation of Apple Mac was bought mainly for graphics applications, whereas the PC was traditionally bought for text-related work. The distinctions have become blurred over the years, and though the PC running Windows is by far the dominant machine, some firms have been replacing PCs by Macs on the grounds of better defenses against viruses (a factor common to all OSs based on Unix).

The Mac OS X is frequently updated by Internet downloads, more often than Windows and probably about as frequently as Linux. The current version is referred to as Snow Leopard, and this reference to animal names seems to be popular with non-Windows OSs.

Mac OS X can be used as a command line system, with commands that will be familiar to anyone who has used Unix or Linux. Most users, however, will make use of the GUI called Aqua.

Linux

Linux started as a freeware version of Unix. From the start of microcomputing there has been a movement that believes that software should be free and widely distributed. In 1983, the GNU project was launched to develop Unix-based software that would be free to users. Linus Torvalds at the University of Helsinki then developed his own version of Unix, using GNU applications.

The Linux OS is unique in the sense that it can be obtained free in source-code form (the instructions in text that can be converted into microprocessor instructions), so that any user with advanced programming skills can modify and distribute a version. This is why so many versions of Linux are available. Not all versions of Linux are free, but most of the applications that run under Linux can be downloaded free from the Internet. The applications may be developed by a small group of programmers or by a large corporation (for example, the Open Office suite was developed by Sun Corporation, and subsequently sold to Oracle Corporation). The fact that so many applications are also available in source-code form means that improvements can be made continually by programmers other than the original developers.

A useful feature of Linux is that versions of it can be installed on a very wide range of hardware, from mobile phones and netbooks to desktops and laptops, and also to mainframe and supercomputers. Web servers in particular are likely to use Linux, and the uptake on PC desktop machines is on the increase thanks in particular to the free versions from Ubuntu.

Linux can be bought on CD, or downloaded to your existing PC and then burned on to a CD or a flash drive for installation. The packaging comprising Linux along with a selections of applications constitutes a Linux distribution (**Distro**), so that by installing a distribution you have a good OS and practically all of the applications that you are likely to need. The cost saving is quite spectacular if you compare this with other OSs and applications. Well-known distributions include Ubuntu (a particularly good choice if you are switching from Windows), OpenSuse, Fedora, and Red Hat.

■ Note

There are free applications (but not OS) for Windows and for Mac OS X as well, but some of these are very much freelance efforts that are not encouraged by the OS manufacturers.

The Ubuntu distribution is, at the time of printing, in version 11.04. The numbering system consists of year, decimal point, and then month digits, so that new versions appear in April and October each year. If you are installing for the first time you can download an installation copy that you burn on to CD and then install. From then on, updates come through the Internet with no need to use a CD.

The desktop version is usually supported for three years and the server version for five years, but you may find that the version that you use needs very little support after three years and unless problems arise you can use it indefinitely. One point to remember is that when you install from CD, this usually scrubs the memory, so that if you had an old version and wanted to install a new version from CD, you need to keep your files and settings by saving the disk folder called 'Home'. An Internet upgrade has no effect on the Home folder contents.

Figure 15.2 shows part of the typical starting screen view for Ubuntu Linux with the other menu portions of Places (folders and drives) and System (change essential features).

One very useful point about the Ubuntu distribution is that the CD allows you to run the OS (slowly!) directly from the CD, so that you can try it out without committing yourself to it until you are happy to use it. This is also a feature that allows you to recover from a disaster that has left your computer locked up. The form of the GUI looks familiar, so that if you have used Windows or Mac OS X you will quickly settle to Ubuntu. The amount of support is impressive: the community of users is always available to discuss problems or misunderstandings, and if you make it clear that you are new to the system you will not be responded to with incomprehensible programmer-speak. All this has made Ubuntu an excellent choice for anyone who has built or upgraded a computer for himself or herself.

■ Note

A Linux installation will (unless you opt otherwise) totally clear your hard drive and reorganize it. There are procedures for twin installations (Windows and Linux), but these are aimed at the more advanced user.

Figure 15.2:
Part of the typical starting screen view for Ubuntu Linux with the other menu portions of Places (folders and drives) and System (change essential features)

■ Note

In 2010 Google announced Chrome, a Linux-based OS that is based on Linux and constructed around the Google Chrome browser. A machine using Chrome will have very few applications on its hard drive, and will download applications as required and for as long as required. This allows small storage capacity to be used, a factor that is important for notebook/notebook and tablet computers. A fast Internet connection is needed.

■

Setting Up

Whatever OS you use, you will need to arrange your hard drive so that you can achieve easy operation and minimum clutter. Every OS will provide some organization for you, but this may not be what you want. For example, you will find a folder called 'Documents' provided by your OS, but do you want to keep all of your documents in this folder, or would you like to have sub-folders called 'Letters', 'Notes', 'Accounts', 'Labels', 'Holidays', and so on?

Installing Applications

Whatever OS you settle on, other than Chrome, you will have to install the applications that you want to use. You have the choice of the well-known and costly varieties (unless you use older versions) or the multitude of freeware that is now available, and you also have the choice of CD or Internet installation. CD installation will always have a direct cost, but remember that if you install from the Internet you may be paying for the download unless you are on unlimited broadband.

Many applications come on a single CD but some, particularly graphics applications, may need more than one CD (or a DVD) and some may come with data files as well as program files. Major applications supplied on CD will come with a manual, but to get the best out of them you may need more. The manual that accompanied the early versions of Microsoft Word, for example, was a treasure trove of information, and there is no modern equivalent unless you are prepared to pay for a third-party book. Similarly, the excellent Paint Shop Pro graphics program used to come with an excellent manual, as did Corel Draw, but now that these names have amalgamated you will have to rely on Internet tutorials or third-party books for all but the most obvious tasks.

Drivers

One class of software that causes more problems than any other is the driver. A driver is a piece of software (which can be quite large) that enables your computer to use a piece of hardware. For example, a printer will need to have a driver in your PC, and if you change to a different make or model of printer you may have to change the driver as well. It might be possible to use the printer for some text documents without a driver, but for anything else you definitely need to have the correct driver installed.

The same is true for your monitor, scanner, WiFi adapter, and any other hardware apart from keyboard, optical drive, and mouse (and even some mice need a driver if they include features that are not part of a bog-standard mouse). A graphics board will almost certainly need a driver if you are to use it for anything other than displaying text.

For the Windows user, drivers should not be too much of a problem because all hardware that needs a driver will include a Windows driver on a CD. You need to be sure that the driver is the correct one for your version of Windows, but unless you are using something prior to Windows Me you are unlikely to have any problems. It is more of a problem if you acquire a piece of hardware that has lost its installation CD.

Though you can download drivers from the Internet, you have to be really careful about specifying the hardware correctly. If you glance at the driver list for HP printers, for example, you will see what I mean. An added complication is that some hardware has drivers from

Microsoft already in your PC but the manufacturer has provided different drivers. The safest way out is to use the Microsoft drivers unless you find that they are really unsuitable (which is unusual).

Life is not quite so easy or inexpensive for Mac users and is much more complicated for some Linux users. New hardware nowadays is likely to provide Mac drivers as well as Windows drivers, though older hardware might not, and you might not be able to find suitable drivers. Drivers for Linux are very often built into your Linux distribution, but you cannot be sure until you check it out. Certainly you should not expect that very new hardware will have a Linux driver when it first goes on sale, but if the hardware is widely used you can be fairly sure that someone in the Linux community will write a driver.

Old hardware can be a pain: I had a very good HP scanner that gave me years of service with successive version of Windows, but no Linux driver was ever available, and I had to solve the problem by looking (on Google) for a new scanner that was Linux compatible (it turned out to be the Canon LIDE, and it has served me really well in my Linux days). If you are building up a computer system for yourself with the intention of using Linux, just check that the hardware you want to use is compatible.

Other Software

When you install a major application a large number of bits of software are added that you know next to nothing about. An example of this is the **codec**. The word is derived from code–decode, and it applies to files of sound or video.

For example, if you record a sound from an analog source into a digital file you need to use a codec to convert analog to digital and to compress the digital signals to fit into memory. The compression can be lossless, meaning that when the reverse codec is used to convert digital to analog for a loudspeaker the resulting analog signal will be as near as possible identical to the original. The alternative is lossy compressing, which can greatly reduce the size of a file but with some reduction of quality (which may or may not be easily detected). Another major use of codecs is conversion from one format of file to another.

Codecs are also used for video files where compression is important because of the size of such files. For example, a piece of video using a codec in the AVI format may be of 22 MB for a short clip, using the MPG codec makes this 17 MB, but if the WMV codec is used the size reduces to 1.4 MB (but the quality may not be adequate).

Codecs are widely available, usually free, for a huge range of purposes, and you can convert almost any form of file of audio, still photo, or video into any other if you have the correct codecs. Once again, these are easily obtained for Windows and Mac (though many websites will try to persuade you to part with cash). Linux users working with video have the luxury of a free piece of software called WinFF, and they can also download a large number of other codecs from the Linux sites.

Skype

Skype is a system that allows you to make telephone (including videophone) calls to other computer users (and also voice-only calls to any telephone subscriber, with or without a computer). You will need a microphone and (if you want to make video calls, a webcam). You must not use Skype for emergency calls (999 or 911). The software for Skype is free, and uses software that uses the voice over Internet protocol (VOIP) standard.

The service was launched in 2003 and in use it resembles an Internet chat room. Making contacts with other Skype users is free, but if you make calls to a telephone landline or mobile phone you have to pay (though the charges are lower than you would encounter using your own landline or mobile). Payment can be made by buying credits from Skype.

CHAPTER 16

Digital Television and Radio

■ Note

Do not read this chapter in isolation. You will need first to have read and understood the material in Chapters 11 and 12.

■

Compression

Digital television and audio signals were in use for studio recording and editing well before broadcasting of such signals became possible. In closed circuits, bandwidth is of little consequence, but broadcasting is bound by the existing frequency bands. What made broadcasting of digital television and radio possible was the development of compression methods, and in particular the MPEG standards.

Chapter 12 has already outlined the compression methods that are used, classed either as lossy or non-lossy. Though non-lossy compression is used for digital signals in a closed (studio) network, it cannot provide enough compression for use with broadcast signals for which bandwidth is limited, and the various lossy systems described earlier along with the techniques of interleaving are also used. Error-detection and correction systems are also important for broadcast signals, and in addition to interleaving and Reed—Solomon methods, another advanced error-correcting system called **Viterbi** coding is also used.

■ Note

Lossy compression must not be used on signals that have to be edited, because after several editing and saving actions the effect of the repeated compression might make the picture unusable. All lossy compression is therefore applied just before transmission.

■

Television Coding

Definition

A digital television signal is created by sampling, either in the camera itself or on an existing analog picture (such as from an analog recording or from a film scanner). The standard digital

Electronics Simplified. DOI: 10.1016/B978-0-08-097063-9.10016-0

format allows for either 525- or 625-line pictures, making it easier to exchange information with US systems (though the sound processing is different), and is also designed to be compatible with high-definition television that is now available both from satellite sources and on free-to-air transmissions such as Freeview.

In this book, we will confine most of the descriptions to the older UK 625-line system. High-definition (HD) transmissions can use either 720 or 1080 lines, but the compression that is used can wipe out some of the gain in resolution. Trials of high-definition television (HDTV) started in 2006, and though the initial program output was available on cable and satellite channels only, HDTV on Freeview has been available in some areas since December 2009. Three-dimensional (3D) television is also available in some areas.

A few television receivers have built-in provision for receiving HDTV satellite or Freeview signals, but the vast majority of domestic receivers are at the time of writing only **HD-ready**, meaning that they need a set-top box to receive the HD channels (which are separate from the normal channels even if they are showing the same programs). In addition, the HD set-top box has to be connected to the HD-ready receiver using a high-definition multimedia interface (HDMI) cable. The connector for this looks at first sight like a computer USB connector, but they are not compatible. The HDMI connector is also used to connect a Blu-ray player to an HDTV, and this provides a full 1080-line image quality (higher resolution than any broadcast signal).

Getting back to 'ordinary' digital television, we might expect from Chapter 8 that the complete analog signal would be sampled and digitized. This, however, is neither necessary nor desirable. For one thing, we do not need sync signals for digital television because synchronization is achieved by clock pulses. The other thing is that the television signal can be divided into luminance (monochrome) and chrominance (color) components. The luminance signal (taking a PAL signal as an example) is sampled at the rate of 13.5 MHz (864 samples in each line), but the chrominance signal is sampled at only 6.75 MHz for each of the two color-difference signals (432 samples in each line, alternately R−Y and B−Y). Using a lower sampling rate for chrominance reduces the data rate and so makes less call on the bandwidth.

■ Note

The sampling rate used for 525 lines is the same, and the frequency of 13.5 MHz was chosen because it is a multiple of 2.25 MHz, the lowest frequency that is a multiple of both 625- and 525-line frequencies.

■

Each sample is 8-bit coded (compare this with the 16-bit coding used for sound), and of the possible 256 levels that 8-bit coding can provide, only 220 are used for luminance, with black corresponding to level 16 and peak white corresponding to level 235. The chrominance coding uses 225 levels, with zero signal corresponding to level 128. This allows for coding both positive and negative value of the color difference signals. Levels 0 and 255 are used only for synchronization.

■ Note

The sampling, using a clock rate of 27 MHz, produces a multiplexed signal, with the form Y-U-Y-V-Y-U-Y-V …, so that each luminance (Y) sample is followed by one of the chrominance (U or V) samples, alternated. The clock synchronizing bits are added at the end of each line, and at the end of each field, and the digitized sound (up to eight stereo channels) can be placed in the line blanking intervals, using about three samples of audio per line blanking interval. The requirement is 3.072 samples per line, so that some lines will use three or fewer, and some will use four.

■

Conversion and sampling are done by an integrated circuit (IC) flash converter. The principle is illustrated in Figure 16.1. A stabilized 2 V supply feeds a chain of 256 (IC) resistors. Each point where two resistors connect is fed to the input of a comparator, for which the other signal input is the analog video signal, set to a maximum amplitude of 2 V. A clock pulse is used to enable the comparators, so that sampling takes place when the clock pulse is high. When the

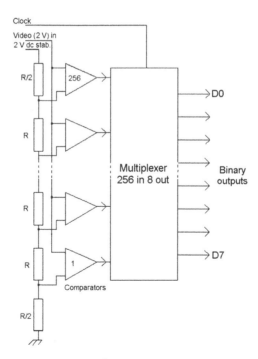

Figure 16.1:
Principles of flash conversion. This is capable of working at very fast clock rates

clock pulse occurs, the number of comparators that have an output will depend on the amplitude of the video signal, and this number is converted to an 8-bit code using a multiplexer.

The video signal has to be compressed, using a mixture of methods, some of which will be varied depending on the nature of the video signal (for example, static picture or fast motion). The compression is such that six channels of digital video (with sound) can be fitted, by multiplexing, in an 8 MHz carrier bandwidth. The multiplexing scheme is based on the principle of a **packet**, as used in optical cable communications systems. Figure 16.2 is a simplified block diagram of the video compression system.

Definition

A packet is a set of digital signals with an identification word or packet identification number (PID) of 13 bits. Each packet is of fixed length, set for television at 188 bytes (or 204 bytes when Reed—Solomon coding is added), and it carries part of the information for a transmitted program. At a receiver, the packets with the chosen PID can be assembled into a stream of data that constituted the signal for a program.

Once data is in packet form, processes such as interleaving, Reed—Solomon coding, and Viterbi coding/decoding can be used to ensure error-free reception. The remaining problem for broadcasting the signal is how to modulate these packets of digital information on to a carrier, and because the three different methods of broadcasting involve quite different problems, three different methods of modulation are used.

Summary

Digital television samples the luminance and chrominance information at different rates, using 8-bit samples. These samples are arranged alternately, and compression is used to reduce the amount of data. The compressed data is arranged in packets, so that each of a set of programs can be identified by its PID. The packet data can be interleaved and coded using Reed—Solomon and Viterbi methods to reduce transmission errors.

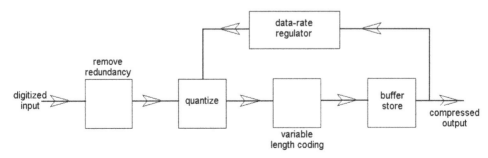

Figure 16.2:
A simplified block diagram for MPEG video compression

Broadcasting Systems

Definition

The broadcasting systems used for digital television are the same as were used for analog television, but the method of modulating the digital signals is different for each transmission system. The three main transmission systems are cable, satellite, and terrestrial transmitter. The differences arise because of the problems that are peculiar to each system. In the lifetime of this book transmission over the Internet will probably become a major way of transmitting digital television and radio. Each type of transmission requires a different set-top box (or internal conversion in the television receiver).

Signal strength is an important factor, because if signal strength is low it may be difficult to achieve a good signal-to-noise (S/N) ratio. On an analog signal, a low S/N ratio will cause the received picture to look grainy, and as the S/N becomes worse, the picture will start to break up. For digital signals, the effect is strikingly different. A digital signal is resistant to corruption until the error correction methods can no longer cope, so that the result, as S/N steadily becomes worse, is that the picture remains unaffected until the S/N reaches a critical level, and then breaks up, so that motion looks jerky and images are pixelated (broken into fragments), and ultimately a complete picture appears only at intervals or the screen displays only a still picture. A low S/N also produces a very objectionable noise on sound. For digital signals, the bit error rate (BER) is a more significant figure than the S/N ratio.

Bandwidth is also a determining feature, because the narrower the bandwidth available the more difficult it is to modulate a digital signal. Using a narrow bandwidth increases the time that the data take to transmit from transmitter to receiver, and if the bandwidth is too small the data will not be received rapidly enough to provide a coherent picture.

Reflections are a problem for any transmitted signal, because if the receiver can pick up reflections of a signal, these reflected signals will act like interfering signals because they will not normally be in phase, and if they are reflected from a moving object their phase will change as the reflecting object (such as an aircraft) moves. Reflections cause the familiar 'ghosting' effect on an analog signal, so that each part of a picture appears with a ghost image displaced horizontally from the corresponding part of the main image. Reflections in a digital image typically have little effect if the reflected signal is at low amplitude, but they will cause complete break-up if the amplitude exceeds a critical value.

Cable

Cable transmission, using fiberoptic cable for all except the last few meters of the signal path, is by far the easiest from the point of view of signal strength, S/N ratio and reflections. The typical cable bandwidth is up to 8 MHz, the S/N is usually high, and reflections are negligible, so that a transmission method of the quadrature amplitude modulation (QAM) type can be used. As the word **quadrature** suggests, it depends on using a two-phase carrier, like the sub-carrier used for chrominance signals in the analog television system.

The two phases can then be modulated independently, and the simplest way to do this is by using each phase to carry one bit, a system called 4-QAM, also called QPSK (see later). With a bandwidth of 8 MHz, however, 4-QAM does not allow the data to be carried rapidly enough, and so each phase of the carrier is modulated at several different amplitude levels. We can use, for example, QAM-16, coding 4 bits for each carrier wave, or we can use QAM-64, coding 6 bits per wave. Using these modulation methods, the data rate can be increased sufficiently to allow successful transmission in the restricted bandwidth.

■ Note

The cable consists of a main line (trunk) that is usually a fiberoptic type, permitting low-loss operation, with converters to radio frequencies at branches that are used to serve sets of individual houses through coaxial cable.

■

At the receiving end, the cable is taken to a junction box at each house, and from there a cable is taken to each socket in the house. A set-top box is needed at each receiver to decode the digital signal from the carrier wave and allow selection of the programs from the carrier. Once the digital signals for a program have been extracted, the normal digital receiver methods are used to convert to the analog video and audio signals. In the early stages of conversion to digital broadcasting, the set-top box will convert directly to analog video and audio signals which are connected to an analog receiver by way of the SCART socket. For digital liquid crystal display (LCD) receivers the set-top box is still required for demodulating the signals, but the connection to the receiver can be digital (such as HDMI).

■ Note

At the time of writing the replacement of old copper telephone lines by fiberoptic cable is planned in the UK. This will allow homes to receive digital television and radio as well as very fast broadband. In cities, cable is already in use, but the replacement of copper by fiber will greatly extend the use of cable for television and broadband.

■

Satellite

The main problem for any transmission by satellite is the very feeble signal that is received on Earth. This is small enough on the dish size that is used for analog reception, and is smaller still on the tiny digital dish. The main problem, then, for satellite reception is the poor S/N ratio, which can be overcome only if the digital signals are modulated in a way that is very resistant to noise corruption. The compensating advantage is that the microwave frequencies used for satellite transmission allow the use of very large bandwidths, typically 30 MHz or more.

The modulation system that is used is termed quadrature phase-shift keying (QPSK) and, as the name suggests, each of two phases can be modulated with one bit, so that using the two phases (at 90° to each other) allows 2 bits to be carried on each carrier wave or set of carrier waves. This type of modulation is much less likely to suffer corruption by noise than any system that codes a greater number of bits per wave. The current microwave band for digital television satellite broadcasting is in the range around 11.7–12.1 GHz.

At the receiving end, the signal is picked up by a small microwave dish, and converted to a lower frequency, typically to a frequency in the range 950 MHz to 1.75 GHz, by a mixer stage, the **low-noise block** (**LNB**), located at the dish. This avoids further losses that would be incurred leading a microwave frequency signal down a long cable. The local oscillator in the LNB is set to a fixed 9750 MHz, so that the selection of a carrier frequency is made at the receiver rather than at the dish. Power to the LNB is provided from the receiver or set-top box through the connecting cable. Figure 16.3 shows the typical block diagram for the LNB. This uses special low-noise transistors made from gallium arsenide rather than from silicon, and the layout of the components is very important because even a few millimeters of conductor can have an appreciable inductive reactance at microwave frequencies.

■ **Note**

The frequency of the local oscillator in the LNB is usually set by a ceramic resonator (altering capacitance) with a screw adjustment. This is placed close to a portion of the circuit but is not electrically connected other than by stray coupling. Dishes can be fitted with twin, quad, or even octo-LNB units so that several receivers can make use of a single dish (or you can use a twin LNB to supply both the television and a separate recorder that has a satellite input, allowing you to watch one program and record another at the same time).

■

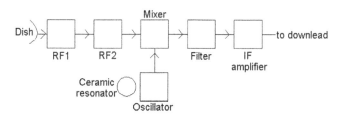

Figure 16.3:
Block diagram of a typical low-noise block (LNB) for satellite reception

At the set-top box, this intermediate frequency is converted to a QPSK digital signal, and then through a set of stages that reverse the coding steps that were done at the transmitter end, eventually being converted to analog video and audio signals.

Terrestrial

Terrestrial broadcasting uses the existing land-based transmitters to broadcast a signal directly to existing antennae. The UK version is called Freeview, and it is similar enough to other schemes to serve as an example. If your antenna (aerial) is capable of receiving an analog signal of good quality, then it should be well up to the task of receiving the digital signals on the same range of ultra-high-frequency (UHF) carrier frequencies. A small fortune was at one time being made in 'Freeview antennae', but if the antenna signal is adequate for analog there is no reason to change it, certainly not before the Freeview signal is available everywhere. Once the UK changeover is complete in June 2011, all transmitters will be broadcasting Freeview at full power, so that it will be only then that the quality of the digital signal can be assessed.

Once again, the UK has chosen a digital broadcasting system that is different from that used in the USA, so making our television receivers more expensive (as we did by selecting PAL rather than NTSC). The difference is that digital signals for the UK are not modulated directly on to one single carrier frequency.

The system that was chosen for digital receivers and set-top boxes in the UK is called **coded orthogonal frequency division multiplexing (COFDM)**. The carrier is coded as if it consisted of a set of 1705 separate carriers, each of 4 kHz, and each modulated, using QAM-64, with part of the digital signal. This makes the transmission rather like a parallel cable with 1705 strands, each strand carrying 6-bit signals. The ICs that code and decode this type of signal are very complex. There was some doubt at the design stage whether the task could be done, and it was for this reason that the 1705 figure was adopted in place of a proposed 8000 carrier system. In fact, the 8000 carrier system could have been used because the technology of manufacturing ICs has progressed just as rapidly as that of the systems using them.

■ Note

The use of COFDM allows a lower S/N ratio to be tolerated, so that digital transmitters can work on lower power without sacrificing the service area. For example, my local transmitter used during the changeover period power outputs of between 7 and 8 kW (effective radiated power) for BBC and ITV, using channel 49 for BBC and 68 for ITV (and C4). The digital transmitters for the same location work at 7.5 kW for channel 39, but at only 1.5 kW for channel 54 and 1.1 kW for channel 50. Power levels will change when the transition to digital is complete.

At the receiver or set-top box, the COFDM signal is reassembled into a continuous stream of digital bits, and is then decoded for program selection in the same way as is used for satellite or cable reception.

■ Note

Even where the normal Freeview signal is strong, there can be momentary interruptions to sound and/or picture. These are much more disturbing than the corresponding faults in analog transmissions and are usually caused by signal echoes. Once the changeover is complete, allowing all transmitters to be run at full power, these problems should ease or disappear.

■

Summary

Digital television signals can be obtained by using an existing antenna and a suitable tuner (which can be in a set-top box or built in to the television receiver). The other option is to use satellite signals (Sky or Freesat in the UK), mainly through a set-top box, but built in to some receivers. Another option is to use cable if you live in an area that has been cabled. In the future, it will probably be possible to receive digital television and radio through an Internet connection once all the copper cabling has been replaced by optical fiber).

Digital Television Receivers

In a digital television receiver, the tuner contains chips that allow for either satellite signals (from the LNB for Sky or Freesat) or antenna signals for a Freeview receiver to be selected and then decoded to a stream of digital signals. Out of this stream you can select the single program that you want to watch by using the PID (program identity) number that is broadcast with the signal. Selecting the appropriate PID provides the digital signals for one particular program. Because cable transmissions are not available outside cities, no receivers are currently provided with cable reception.

Once this digital stream has been selected, the coding actions are reversed so that the original signals can be retrieved. This set of actions produces an 8-bit luminance signal and two 8-bit chrominance signals together with the digital audio signal. These signals are converted to analog for use in the conventional audio and video sections of the receiver.

DVD

DVD, originally meaning digital video disc, is now taken to mean **digital versatile disc**, and it refers to a development of CD technology that is now fully established (and likely to be superseded by Blu-ray; see later). This was originally directed to recording full-length films on CD, hence the 'video' in the original title, but the idea has been extended to a universal type of storage disc that can be used for films, audio, or computer data interchangeably. Inevitably, as soon as writable and rewritable DVDs could be manufactured (working along the same principles as CD-R and CD-RW), writing drives became available for computers and are now universally fitted except for the sub-£100 group of notepad computers that are now available.

■ Note

An important feature of a modern DVD computer or television drive unit is that it will accept conventional CDs as well as DVDs.

■

The DVD holds much more data, can transfer it more quickly, and is as easy to reproduce by stamping processes as the older CDs. However, by the time DVD became the uniform recording format, replacing cassettes, DAT, videotape, and CD-ROM, it was already close to becoming outdated. In the USA, where Blu-ray discs are relatively cheap, this format has to a considerable extent replaced DVD. In the UK, however, Blu-ray has not taken off as rapidly as was expected.

Film rentals are the main end-use of DVD at present, and video-cassettes are now so out of date that charity shops are flooded with them. Surveys have shown repeatedly that the most common use for video-cassette recorders (VCRs) in the UK is to record television programs either when the viewer is not at home, or when two interesting programs are being broadcast at the same time (yes, it can happen). Though recordable DVD machines are widely available now they have never achieved quite the level of sales that VCRs did, though they offer a much more user-friendly option (no more recording over a show that you wanted to keep, or searching all over a tape to find what you want to see). Even this use of DVD, however, is likely to pass away now that so many set-top boxes incorporate a hard drive (and some also add HD).

For computers, DVD offers so much more storage space than CD that the options it allows are more than most users can cope with at first. A single-layer disc can store 4.7 Gbytes, a large chunk of data, corresponding to just over two hours of digital video signals at a higher quality than is possible using VCR (which relies on considerable bandwidth reduction). More than one layer of CD recording can be placed on a disc, however, because the layers are transparent, and by altering the focus of the reading laser, it is not technically difficult to read either of two superimposed layers that are only a fraction of a millimeter apart.

By making two-layer DVDs the recording time can be doubled, and by adding double-sided recording it can be doubled again to eight hours of video. The discs can contain up to eight audio tracks, each using up to eight channels, so that films can contain soundtracks in more than one language, and cater for surround-sound systems.

The DVD can also end the concept of a film as a single story, because unlike tape it can switch from one set of tracks to another very quickly, allowing films to be recorded with several optional endings, for example. Shots taken using different camera angles can also be selected by the viewer from the set recorded on the disc, and displays of text, in more than one language, can be used for audio and video tracks. Like CD and so unlike VCR, winding and rewinding are obsolete concepts, and a DVD can be searched at a very high speed that seems instantaneous compared to VCR. The disc is also smaller than a video-cassette, does not wear out from being played many times, and resists damage from magnets or heat.

■ **Note**

DVD for video uses MPEG-2 coding and decoding. Even MPEG-2, however, is a lossy compression method, and this sometimes shows in video quality as shimmering, fuzzy detail, and other effects.

Summary

DVD is a development of CD, capable of using both sides (and possibly two layers per side) so as to increase the storage capacity, together with a short-wavelength laser to allow tighter packing of data. This allows for several hours of compressed video to be stored on a disc. Writable and rewritable DVDs are available at low prices both for video recording and for computer data storage.

Digital Radio

We have seen from this and earlier chapters how analog sound signals can be sampled and converted into a stream of digital bits, together with the coding and interleaving methods that can be used to avoid errors on replay. The CD was the first digital sound consumer product, and the enormous developments that have taken place in compression methods since then make more modern systems such as MP3 and digital radio possible.

■ **Note**

There is no single system for digital radio, and the world seems to have divided into two regions, one mainly in European and Commonwealth countries, using a system called digital audio broadcasting (DAB), and the other (in the USA) using the in-band on-channel (IBOC) system on FM frequencies, HD on other frequencies, and also digital satellite radio. At the time of writing, 30 or so countries are committed to DAB, and many are testing or committed to IBOC.

Digital radio, unlike digital television, does not offer much to anyone who is satisfied with the existing system. If you have good FM reception of your favorite stations at home, then a change to digital will make a large difference to your bank account but very little change to your ears. Inevitably, the FM broadcasts will be closed down in the UK, because the UK Government has sold the rights to these frequencies in advance to the mobile phone companies (they can help in extending the mobile phone network by making less use of landlines for long-distance hops). It seems at the time of writing that some FM frequencies will remain in use, particularly for local radio.

There has been a lot of questionable advertising for digital radio, claiming better sound, but since the sound (for most of the digital radios on offer) comes from a small loudspeaker it is hardly likely to be any better than an FM radio with the same size of loudspeaker. If pressed, the advertisers will admit that they are comparing digital radio with medium-wave broadcasts (now that everyone listens to FM). Audio devotees, who are likely to spend several thousands of pounds on loudspeakers alone, reject the idea of digital radios being superior to FM, mainly because of the effects of compression.

The most serious problem is that of reception areas. Digital radio broadcasting in the UK uses part of the old band III television frequencies, 217.5–230 MHz, and the coverage for these frequencies was originally plotted for the old ITV signals, for which everyone used an outside rooftop antenna. Today's digital portable radios use a whip antenna, and the signal strength is considerably less than that of an external yagi. Not surprisingly, reception for digital is patchy, and this is not exactly endearing for consumers who find that they can receive their favorite station only if the radio is placed on an upstairs bedroom window-ledge. US listeners, by contrast, can receive digital radio by IBOC or by satellite.

■ Note

One other problem that has plagued portable DAB radios is power consumption. Older DAB radios had a very short battery life because of the power-hungry DAB chip. Modern units have reduced the power demand, but the chips need around 4.5 W of power, so that many DAB portables use very small loudspeakers (hence the poor sound quality of so many). The best compromise seems to be a power-efficient chip combined with a lithium-ion battery.

At one time, the main advantage was seen to be for users of car radios, because at present if you want to listen to a national station on FM, you have to retune several times in the course of a long journey (though the present generation of FM car receivers can do the retuning automatically). Digital radio, in theory, obviates the need to retune, because all the transmitters for one national station can use the same frequency rather than the different frequencies used at present to prevent one station interfering with another. In practice, as one observer has noted, the main use for digital radio in a car is to find where a usable digital signal might be found.

This has not prevented the government leaning on car manufacturers (who know what customers prefer) to install digital radios in all new cars. The time scale that has been mentioned for the UK is 2015, with the assumption that 50% of all radios will be digital by 2013. At the time of writing, in-car radios that use DAB have FM tuners also, and will switch automatically to FM when (not if) there is no adequate digital radio signal. Digital radio can be received on some motorways and near cities, but reception is patchy over large areas of countryside, and since there is not likely to be a massive program of building new transmitters it looks as if the 50% target will remain a dream for many years to come.

The signals themselves use the same COFDM type of system as was described for Freeview television, but with 1536 carriers for each channel. This allows for multiplexing several signals on to a single data stream. One particular advantage is that the COFDM system is so resistant to interference from distant transmitters that one central carrier frequency can be used for several transmitters (for national broadcasts), ending the misery of trying to listen to a national station going in and out of tune in a moving car. Once coverage has extended to 90% of the country (as has been assumed in various surveys) this ambition can be realized, but it still leaves questions, such as what will be done to listeners in areas where the DAB signal is just about adequate for the BBC channels but no others?

IBOC

IBOC is a digital radio broadcasting system that makes use of the sidebands of FM (or even AM) transmissions to send out digital signals. This has arisen particularly in the USA because the higher frequencies that are used for DAB in Europe are not available in the USA. There are three varieties of IBOC, of which only one, HD Radio, is at the time of writing approved for use. The other two versions are FMeXtra and DRM+, which differ in the way that the additional digital transmissions are modulated on to existing FM or AM carriers.

The take-up for IBOC by radio listeners has not been spectacular. One estimate has put it as low as 1%, compared with the 30% of UK listeners who can and do receive DAB.

Satellite Digital Radio

Satellite broadcasters around the world include radio channels as well as television channels, and set-top boxes for satellite reception usually have audio output sockets that can be used to link them to audio equipment. Sound quality can be excellent (though this depends on the source of the signals). Dedicated satellite radio receivers are rare in Europe but are much more common in the USA, where there has been a strong take-up of digital satellite receivers in cars.

CHAPTER 17

Miscellaneous Systems

This chapter is concerned with some items that were omitted earlier, either because they did not match the content of other chapters or because you might have had some difficulties in understanding them until other chapters had been read.

Radar

Classic radar principles are similar in many ways to those used for television. Radar depends on the fact that the speed of electromagnetic waves in air is constant, identical to the speed of light at 300 meters per microsecond. If radio waves are sent out as a beam and are reflected from a target, the reflected wave will be detected some time after the transmitted wave, and the time delay will be equal to the time for the waves to travel to the target and back again. For example, a target at 300 m distance will correspond to a delay time of 2 µs, and a target at 30 km distance will cause a return beam (an echo) to appear 200 µs later.

The classical type of radar that has been used for more than 50 years is the **plan-position indicator** (PPI) which produces on a cathode-ray tube (CRT) (or other display) a pattern in which the base station is represented by the center of the screen and the targets appear as bright spots. The distance of a target dot from the center of the screen is proportional to the distance of the target from the transmitter, and the bearing of the target is given by the angle from a zero degree line drawn on the screen (or on a cursor that fits over the screen) (Figure 17.1). Though there have been considerable developments in radar systems over the past 50 years, these basic principles remain almost unchanged.

The requirements for radar are the transmission of short pulses of very high-frequency waves, the demodulation and amplification of the return signal, and the use of the return signal to display on a CRT or, more usually now, on a flat panel after digital processing that allows other information (such as identification codes) to be displayed. Though lower frequencies were used initially, the frequencies that are used are nowadays several gigahertz, and they are generated by vacuum devices called **magnetrons**. The magnetron is a form of oscillator which causes radio wave oscillations in much the same way as a flute player causes sound oscillations. The magnetron uses a magnet to make an electron beam take a circular path, and by steering this path across circular cavities in a block of metal, the beam can be made to oscillate at a frequency that is determined by the dimensions of the cavity. The cavity will typically be of a few millimeters radius, so that the waves that are generated have a wavelength of a few

Electronics Simplified. DOI: 10.1016/B978-0-08-097063-9.10017-2

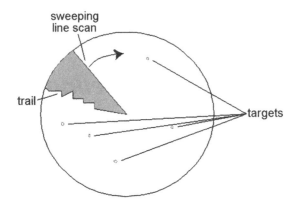

Figure 17.1:
The appearance of a traditional plan-position indicator (PPI) display, showing the rotating scan. The tube is always of a long-persistence type, so that the scan line leaves a trail, and targets appear as slowly fading blobs. The range is indicated by the distance from the center of the screen, and the bearing is indicated by the angle of the target blobs

millimeters, corresponding to frequencies of several gigahertz. These frequencies are called the microwave range, and the familiar microwave oven uses a magnetron working at 2.45 GHz, which is a frequency that matches the natural frequency of vibration of molecules of water. Magnetrons can put out pulses of very high energy and no semiconductor substitute has been found, though semiconductor devices can generate microwaves of low power.

Summary

Radar makes use of waves in the gigahertz frequency range. These have a short wavelength, behave like light waves, and reflect from dense objects. Specialized vacuum tubes called magnetrons can generate these waves, usually in short bursts. When a burst of microwaves is beamed from an antenna and hits a target, the echo will return in a time that depends on the distance of the target.

The classic type of magnetron is pulsed by switching the power on and off, typically for one microsecond or less in each millisecond. Since the magnetron is switched on for only one-thousandth of the time when it is being used, the amount of power in each pulse can be very large without causing cooling problems. For example, if the power in a pulse is 1 MW (one million watts), the average power might be just 1 kW (1000 watts), small enough to cope with by air or water cooling of the magnetron.

Figure 17.2 illustrates a block diagram for a PPI radar station. The clock circuits generate the pulses, setting the time between pulses (this is more usually expressed as a frequency, the **pulse repetition rate** or **PRF**). One output from the clock is used to generate high-voltage pulses of

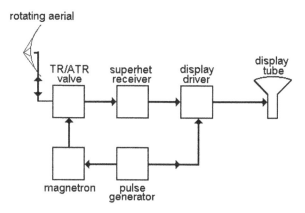

Figure 17.2:
A block diagram for a plan-position indicator (PPI) type of radar system

the duration that is required, and these pulses are applied to the magnetron by way of a modulator. The pulse of microwaves passes through the TR/ATR tube, which separates the transmitting circuits from the receiving circuits. This is an essential part of any radar in which a single antenna is used for both transmission and reception, because with typical transmitter pulse power of 1 MW and typical received signal of a few microwatts, the receiver would be vaporized if the transmitter signal reached it.

The antenna consists of a dish similar to the type that is so familiar for satellite reception (which typically uses around 11 GHz frequencies). Radar dishes are usually larger because space is not usually a problem, and frequencies lower than 11 GHz are being used. In addition, the dish rotates so that it aims its microwave beam in a slightly different direction for each pulse. When the pulse of microwaves has been transmitted, the TR/ATR tubes close off the transmitter circuits from the antenna and open the receiver circuits.

The returning echo is received on the dish and its signal is passed to the receiver which, as usual, is a superhet design. At the demodulator, the signal consists of a pulse identical in width to the original pulse. This signal is then gated by another pulse that is derived from the clock circuit. The purpose of this is to keep unwanted signals from the display by allowing signals to pass only for a specified time after sending out a pulse.

The classic type of display consists of the CRT with a scan that is radial, deflecting the beam from the center of the tube out to the edges. At one time, the scanning coils were rotated in step with the rotation of the antenna dish, but by the 1950s it was more usual to keep the scanning coils steady and alter the phasing of the scanning signals so that the direction of deflection of the beam rotates in step with the rotating antenna dish. The received pulse is used to brighten the beam, providing a target display, and a graticule (a transparent scale) can be used to read direction and distance.

On modern radar displays, computer techniques place a reference number against each target so as to make it easier to follow the movement of a target, and figures for the size, distance, bearing, and speed of a target can be obtained from the computer system. Modern radar uses a raster form of display (more like a television receiver, and using a flat screen) but showing the information in PPI format. Computer technology can be used in such a display to show markers so as to identify the type of target.

Doppler Radar

Doppler radar is used for height and speed measurement rather than position, and it depends on a continuous lower power microwave output rather than the use of pulses. On a Doppler system, microwaves are sent out and received continuously, and the two waves are mixed together, using circuits that ensure that both are of the same amplitude. For a target that is not moving, the returning waves have exactly the same frequency as the transmitted waves, but when a target is moving the returning wave has a slightly different frequency. This returning frequency is higher if the target is moving towards the antenna, and is lower if the target is moving away from the antenna.

The difference between the transmitted and the returned frequency can be used to measure the speed of the target to or from the antenna, and on modern equipment, a computer is used to work out speed. For aircraft height measurements, the antenna points vertically downwards and the phase of the signal returned from the ground is compared with the phase of the transmitted signal, and the difference used to find the height. This is much more precise than the barometer type of altimeters that were in use into the 1950s even on military aircraft.

Summary

The PPI type of radar uses a rotating antenna and a display that consists of a scan line rotating in synchronism. The returning echo pulse is amplified and used to brighten the trace, so that the distance from the center of the CRT represents distance and the bearing of the trace is the same as the bearing of the target. Doppler radar uses continuous rather than pulsed waves and measures the speed of a target from the change of frequency of the echo signal. Phase changes can be used to measure distance, notably small distances such as are required for height measurements.

The Oscilloscope

Definition

The oscilloscope is an instrument that displays waveforms, allowing measurements of wave time and amplitude to be taken. Older oscilloscopes are analog instruments, but digital oscilloscopes are now available, permitting more elaborate analysis of a waveform. We will look at the analog principles first, because there are several million analog oscilloscopes around and they are still useful instruments. Because the oscilloscope display closely resembles a graph, we use the terms X and Y, with the X-axis meaning the horizontal line and Y-axis meaning the vertical.

This description assumes that you have read the description of the cathode-ray tube in Chapter 8, and it refers to the older analog type of oscilloscope. The block diagram of the electronic circuits of a simple oscilloscope, omitting the CRT, power supply, and brightness controls, is shown in Figure 17.3.

A signal applied to the Y-input is attenuated by a calibrated variable attenuator (usually a switch rather than a potentiometer) before being applied to the Y-amplifier. The amplified output drives the Y-plates, which cause the spot to be deflected in the vertical direction. The gain of the amplifier and the settings of the attenuator are matched to the sensitivity of the CRT, and the attenuator is consequently calibrated in volts/cm, so that the amplitude of a waveform can easily be read from the screen.

Some of the Y-input signal is fed to a synchronizing circuit which generates a pulse at every signal peak (either positive or negative peaks, usually selected by a switch). Each of these synchronizing pulses is then used to start a **timebase**, a sawtooth signal that is applied by the X-amplifier to the X-plates.

The effect of such a sawtooth waveform is to deflect the electron beam and so the spot, at a steady speed across the screen in what is called a **sweep**, and then to return it very rapidly to its starting point in what is called **flyback**. The speed of the timebase sweep can be controlled by a calibrated switch, so that the time needed for the spot to scan every centimeter of the screen can be printed on the switch plate. From the user's point of view, the timebase sweeps the spot across the screen from left to right. The sweep can be continuous or it can be **triggered** when the input signal reaches a set amplitude.

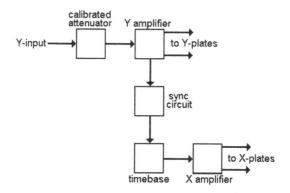

Figure 17.3:
Principles of the oscilloscope, the essential instrument for working with waveforms

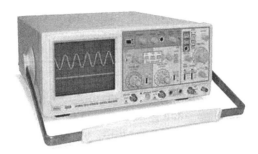

Figure 17.4:
A typical traditional oscilloscope and its trace

Another switch enables the X-amplifier to be used to handle other signals, if desired. When the CRT is in use, a signal of unknown frequency and amplitude is applied to the Y-input. The Y-attenuator and the timebase speed controls should then be adjusted until a steady display of measurable size appears on the screen. The peak-to-peak amplitude of the waveform is then found by measuring, with the aid of the calibration of the attenuator, the vertical distance between peaks on the graticule, a transparent calibrated sheet covering the screen. The time-period of one cycle (τ) is measured on the same graticule, using this time for the calibration of the timebase. The frequency of the signal can then be calculated by the formula: $f = 1/\tau$. Figure 17.4 shows a typical classical oscilloscope trace appearance.

To allow for a wide range of input signals, **probes** can be used. A typical probe would be connected to the Y-input and would use a compensated attenuator at the probe end, so that losses in the connecting cable can be greatly reduced.

The classical analog oscilloscope used one single electron beam, but for many purposes multibeam types (particularly double-beam) are useful. The two beams (which can be produced by a CRT with two electron guns) are deflected by the same timebase signals, but have separate Y-amplifiers and deflection plates so that two signals can be compared using the same time-scale. This is particularly useful for comparing phases and also the timing of pulses.

Summary

The oscilloscope is used for electronics measurements, particularly for waveform measurements on analog circuits. The principle is that a CRT beam is deflected horizontally by a sawtooth waveform, the timebase, while the vertical deflection is caused by an input signal. If the timebase speed is correctly matched to the frequency of the incoming waveform, one or more cycles of the wave will be displayed, and a measuring graticule allows voltage and time measurements to be carried out.

Digital Oscilloscopes and Logic Analyzers

The **digital oscilloscope** that has replaced the analog type for commercial users is usually of the storage type (**DSO**) because storage is a particularly valuable feature that was much more difficult to implement using purely analog methods. On a digital oscilloscope the data can be held in memory for as long as is needed. The logic analyzer is a development of digital oscilloscope that is intended to capture and analyze digital data in a working system, as distinct from displaying analog waveforms.

The basis of the DSO is its input circuitry that has to carry out the conversion of the input analog signal into a digital signal, using an analog to digital (A/D) converter. The data for one line of a trace is then stored in memory under the control of a microprocessor, and can be displayed on a flat liquid crystal display (LCD) panel (replacing the heavy and fragile CRT). Another advantage is that color displays can be used so that traces from different sources can be displayed in contrasting colors. Making a permanent record of a trace formerly had to be accomplished on an analog oscilloscope by using a specialized camera, but on the digital oscilloscope a printer output (as for a computer) can be used to give a paper image. The ultimate advantage is that the DSO can provide much more information about a signal (rise time of a pulse, frequency spectrum, etc.) than would be possible using purely analog methods.

■ **Note**

The combination of an interface card along with software allows a computer to be used as an oscilloscope. Modern fast computers with clock speeds of 1.50 GHz or more are suitable for working with high-frequency signals if the A/D conversion is fast enough.

■

NICAM

Definition

NICAM stereo sound was available on analog television receivers in the UK, and is a method of broadcasting and receiving high-quality stereo sound for television without requiring additional bandwidth. Though the system is not used for digital transmissions, the technology is interesting, and it provided methods that were harnessed for use in digital signals.

The stereo broadcasting system that was used for frequency modulation (FM) was not suitable for analog television because the additional bandwidth of the sound channel would overlap the television bandwidth, so that a different method had to be used. NICAM is an acronym for near instantaneous companded audio multiplex, and as the name suggests, it uses the digital methods of conveying two-channel signals by sending portions of the two channels alternately, as is used also for compact discs (CDs).

All NICAM systems retain the normal sound carrier signal at 6 MHz above the vision carrier and carrying the mono signal, so that there is no compatibility problem for receivers that do not have the NICAM chips. At the transmitter, the two audio channels are sampled alternately at a rate of 32 kHz and with 14-bit digital output. This is then compressed to 10 bits, the digital equivalent of a type of Dolby process, and the usual parity bits are added.

The signals are then interleaved, as is done for CD (see Chapter 12), and assembled into 728-bit 'frames'. The digital signals are then phase-shift modulated on to a carrier which is at 6.552 MHz above the vision carrier. At the receiver these signals are separated from the FM mono sound signals (by filtering), demodulated, and separated into left and right channels using a storage chip so that both channels can be fed simultaneously once again. Finally, the 10-bit signals are expanded to 14 bits again and converted to analog form as the left and right audio outputs.

Summary

NICAM is a method of transmitting high-quality stereo sound that was used for analog television receivers, using another sub-carrier. The sound is sampled and quantized, and the digital stream is compressed and coded to make error-correction possible. The system is not used for terrestrial or satellite digital television.

Camcorders

Definition

A camcorder is a combination of a television camera, a video recorder, and a synchronization pulse generator, allowing you to record video and sound on a miniature cassette, digital versatile disc (DVD), or memory card, and subsequently transfer the signals to a full-size DVD or other medium.

Modern camcorders, all digital types, are a triumph of miniaturization, integrated circuit (IC) design, and design ingenuity. Though the picture quality does not match up to what can be attained using full-size professional equipment, the difference is nothing like as great as the price and size difference would suggest, and the poor-quality pictures that are sometimes demonstrated are due more to the failings of the user than to the camcorder. Initially, all camcorders used analog methods, but these have been rapidly replaced by the digital camcorders that are smaller and allow much better picture, better sound quality, easier editing, and longer recording time. Most designs use viewfinders that incorporate an LCD color display, usually on a folding window.

■ Note

Digital camcorders have developed very rapidly, and the choice that is now available is extensive. You can buy camcorders that record on DVD mini-discs, some that record on to memory cards, some that record on a miniature hard drive (even some that still use tape), and even high-definition (HD) models that can record on any of these media or on miniaturized Blu-ray discs.

Camcorders are too small to use camera tubes, even the smallest types, so they employ a form of light-sensitive chip that was originally developed for the television cameras used in space exploration.

Because of the compression, storage space is very efficiently used, and the digital system allows tricks such as recording stills along with some sound. Editing is also possible if you can download the digital data from the camera tape to a computer, work on it, and then re-store on the camera storage system, save it to a file on the computer, or store it to a DVD or Blu-ray disc. On camcorders using tape or memory card this once required a fast signal transfer system called **Firewire** to be fitted on both the camera and the computer. Nowadays it is more usual to use USB2 if transfer is needed; if the camcorder records on to a miniature DVD or to a memory card, no other transfer method is needed as either form of storage can be removed from the camcorder and inserted into the appropriate reader of a computer.

■ Note

Modern still cameras usually provide for a **video mode** which can be quite effective, though it often imposes restrictions such as being unable to zoom while filming in video mode. A more serious problem is that video mode usually means a frame rate of 30 frames per second (f.p.s.), suited to the US NTSC television system. When such video is edited to place on to a DVD for use with PAL receivers, the frame rate is changed to 25 f.p.s. by omitting one frame in six, causing motion to appear jerky on the replayed DVD. At the time of writing only Canon and Panasonic offer 25 f.p.s. on their HD-capable still cameras.

Cellular Phones

Definition

A **cellular phone** (mobile) uses the principle of a low-power transmitter/receiver which can communicate with a base station. The base stations are linked to each other using land lines or microwave networks, so that communication along a long path requires a large number of links.

The first cellular phones used analog methods and were more brick-like than mobile; they were more useful as car phones than for truly handheld mobile use. In a remarkably short space of time, these have been replaced by digital phones, and the old analog network has been shut down. The analog mobile phone (now known as 1G, first generation) has probably been one of the shortest lasting technologies in our time. Even the digital phones have gone through generations 2 and 3 by now. Each generation has introduced new capabilities such as connection to the Internet, e-mailing, photography, and MP3 storage and playing.

Modern cellular phones of the 2G type are constructed to a set of standards known as GSM (Global System for Communication) that started in Europe in 1991. They use the now-familiar ideas of A/D conversion and data compression. Since high-quality sound is not needed, a lower sampling rate and a smaller number of bits per sample can be used. The sampling rate that has been adopted is 8 kHz (allowing for voice frequencies up to 4 kHz), and each sample uses 13 bits initially, though this is compressed later. The compression system reduces the bit rate to 13.4 kb/s (alternatively, the compression can be increased so that the final bit-rate is 6.7 kb/s). The system provides also for a digital channel for text messages using 9.6 kb/s.

GSM phones can use any of three frequency bands, 900, 1800, and 1900 MHz, so that a phone with dual-band or tri-band capability can be used almost anywhere in the world. For any of these bands, some frequencies will be reserved for **uplinking** (mobile to base station) and others for **downlinking** (base to mobile), and the carriers can be used to provide the theoretical maximum of 992 simultaneous conversations. At a base station, the power output is around 20 W, and the power output of a hand phone is usually around 0.8 W. The error-correcting methods that are used can be very effective, making a hand phone usable in a car at speeds of up to 240 kph (150 mph). The modulation system is a form of frequency shift keying called Gaussian minimum shift keying (GMSK).

■ Note

Compare these figures with a microwave oven, which operates at a frequency of 2.45 GHz and a power of typically 700 W. You cannot cook your ears with a mobile phone (though you can get a headache listening to chatter all day long).

■

The main problem with the modern mobile phone is that where you most need communication (such as after an accident on a country road) you are liable to find that there is no usable signal. Coverage varies from one network to another, but is always better in urban areas than in country areas. In some parts of Scotland it is easier to find a public phone box than a mobile signal.

A modern mobile contains a memory card that permits storage of music and pictures, but the important feature for the telephone's features is the **SIM card**. SIM means subscriber identity module, and it is a miniature memory card (usually 25 mm × 15 mm) that stores a digital ID (the service identity subscriber key) that reveals the user's identity to a base station. The SIM card contains data that you have put in place (such as names and numbers of contacts) and also controls what services can be accessed. Two passwords are used, one for low-level protection and one for higher-level protection. When you change to a new phone (assuming you do not want to use any new services) you can remove the SIM card from your old phone and insert it into the new phone. It is also possible to copy your personal information from an old SIM card to a new card that provides many more new features.

3G, the third generation of mobile phone technology, expands the range of capabilities of the mobile, subject only to whatever restrictions are imposed by the network provider. Typical capabilities include mobile Internet, video links, and music downloads. Even older mobiles can now be used for Internet (including e-mail) and MP3 downloads if the network permits (the SIM card will have to be changed).

Cordless Phones

Cordless phones make use of the technology of mobiles to provide a home telephone system with the advantage of using a land line (no more 'no signal' messages). The handset unit is closely modeled on mobile technology, with an address list of your favorite contacts (which can range from ten to 100 names and numbers depending on the make and model of phone). The handset will use either lithium-ion (Li-ion) or nickel—metal hydride (NiMH) cells; the older nickel—cadmium (NiCd) cells are now obsolete for this purpose.

The base unit is connected to your landline and to a power socket, and on several models of cordless phone the base unit can act also as an answering machine. The standard for cordless phone is **DECT** (digital enhanced cordless telecommunications) and the other abbreviation you will come across is **GAP** (generic access profile), meaning that any handset conforming to this standard should work with any base unit that also conforms. This applies to calls in and out, rather than to the use of other features (such as answering machine).

■ Note

If you use computer broadband each of your telephone sockets must be fitted with a filter to separate the telephone signals from the broadband signals (see Chapter 14).

For a typical unit of recent construction, the frequency range is the European standard of 1.88—1.90 GHz and the transmission power is around 10 mW, providing enough range for a house, though reception is likely to be poor at the end of a long garden.

■ Note

Other units use other frequency ranges, including 900 MHz, 2.4 GHz, and 5.8 GHz, and the preferred frequency is really immaterial provided it does not conflict with other radio-enabled equipment. If dialing on your cordless phone causes your neighbor's garage door to open or his car to lock, then one of you will have to change frequency.

■

A feature that is common on modern cordless units is the provision of **speaker-phone**, allowing more than one person to hear an incoming call. Another desirable feature is **caller ID**, but this has to be set up with your phone provider at an extra charge. Caller ID will show the name of any caller whose details are held in the phone address book, and others will be listed by number or by messages such as 'Private Caller'. The latter message usually means that the caller has used a code (such as 141) ahead of your number when dialing you so that no ID is sent (even if the caller is in your address book). Note that ID is pointless if your handset uses a low-resolution display, since the name will usually be unreadable. Other features such as automatic redial, muting, and call waiting are also found on conventional phones.

Bluetooth

Definition

The name sounds curious, and is due to the development of this system by Ericsson of Sweden, who named it after a tenth century Viking king. The modern version is a short-range wireless system that can be applied over a very wide range of uses to allow communication over short distances, replacing cables.

Because the technology is standardized, one Bluetooth device can always communicate with any other that is in range. Both data and voice messages can be handled together, and each Bluetooth in range automatically becomes part of a network of up to seven others. In addition, a Bluetooth-equipped device can belong to more than one network (technically known as **piconet**).

When you see a van driver with a headset, you can be fairly certain that the headset is Bluetooth enabled to allow communication (by Bluetooth-enabled mobile phone) with head office, and that the portable data collection device in his hand is also sending and receiving data.

The technical specification calls for a minimum range of 10 m, but manufacturers are free to implement longer ranges. The frequency range covers 2.4–2.485 GHz with carriers spaced out at 1 MHz intervals, and the Bluetooth technology allows for detecting the frequencies used by other devices and avoiding these frequencies. Power is typically 2.5 mW, and devices will shut down automatically when not in use to minimize energy consumption. Data rates range from 1 to 3 Mb/s depending on the version of Bluetooth device.

The versatility of Bluetooth technology is its most appealing feature. Computer users can use Bluetooth to network between computer and for using a printer, mouse, and keyboard between computers. Bluetooth can also replace the traditional type of serial connections in commercial and medical applications

GPS

Definition

GPS means Global Positioning System, and it refers to a way of locating position on Earth by using signals from two or more satellites.

The system makes use of 24 satellites that orbit the earth in predictable paths at around 177,000 km (11,000 miles) above the Earth. This number of satellites, and the orbits chosen, ensure that at least four satellites (usually eight) will be within the line of sight for a receiver anywhere on Earth. By using information from three satellites, the latitude and longitude position of a receiver can be established, and by using a fourth satellite, altitude can be measured as well. The receiver must find at least three satellites and measure the distance from each.

■ Note

Satellite positions are obtained from a database (an almanac) in the memory of each receiver. Any variation in the position of a satellite will be signaled to each receiver from the US Defense Department base in Colorado, so that the database accuracy is maintained.

■

The distance between the receiver and each satellite is calculated by measuring the time that the radio waves, using a frequency of 1575.42 MHz, take to travel between the satellite and the receiver. Ideally, this measurement would be carried out by sending a coded signal for the time as indicated by a very precise atomic clock on the satellite, and comparing this with the time indicated on an equally precise clock on the receiver. In practice, for non-military applications, it is possible to use a conventional quartz clock on the receiver.

The quartz clock, though good enough for most ordinary applications, does not maintain time with sufficient precision, but it can be adjusted. The principle of adjustment is to calculate the distances from each of four satellites. If the clock at the receiver is not perfectly synchronized with the clocks at the satellites, the distance figures will be incorrect, so that no single point can be found. By adjusting the distance figures so that all the lines between satellites and receiver meet at a point, the time of the quartz clock can be adjusted so that it is synchronized to the atomic clocks, and the single point is the position of the receiver.

Originally, the system was intended only for military use, and its precision was 22 m horizontally and 27.7 m vertically. A civilian version was limited to 100 m horizontal and 156 m vertical. More recently, the higher precision has become available on civilian GPS receivers. Though handheld GPS devices are still available, the main GPS civilian use is for car satellite navigation (satnav). On the whole, this works well, but there are still anomalies that lead drivers to river beds, deep fords, and roads that are too narrow for their vehicles. The combination of satnav and an old-fashioned map is still by far the best way of making an unfamiliar journey. The satnav really comes into its own when you are trying to find a city address on a dark night.

Remote Controls

Remote controls at one time were provided only for the more expensive brands of television receivers, but are now almost universal for television, radio, DVD players or recorders, set-top boxes, and such other purposes as garage doors, cameras, car door locks, electric and gas heater controls, and such hobby items as model aircraft.

It is also possible to buy socket adapters that can be placed between a mains socket and an appliance plug to allow any electrical appliance to be switched on and off remotely. This latter use is very appealing to anyone who likes to believe that a vast amount will be saved by switching off appliances rather then leaving them on standby. Early types of television remote control used ultrasonic signals, but these are obsolete now.

The remote controls for television, radio, DVD, and electric or gas heaters are of the **infrared** (**IR**) variety, but the others (for use outside the house) operate using radio transmitter and receiver technology. These latter types are liable to problems of interference, and there are well-documented examples of such problems (such as the street where all parked cars unlocked when a shop till was activated). Radio remotes can offer much greater range (typically 0.8–3.2 km; 0.5–2 miles) and are not seriously blocked by buildings in the line of sight.

Taking the television type of remote control first, these operate by using a control system to send digital codes along an infrared beam from a light-emitting diode (LED). The binary code is carried using a brief burst of IR followed by a 'dead' time (no transmission) for each bit. The precise coding depends on the manufacturer of the equipment, but the different systems are converging, so that it is possible to adapt a remote intended for a DVD player to control the television as well, and there are also 'universal' IR controllers that can be programmed to any action that can be provided by IR. The snag with these universal remotes is that you have to specify what unit is being controlled before you send the command. The effect of this is that universals work best when you have only a few devices to control. All IR units have a fairly short range, and are effective only if there is a line of sight between the remote and the device being controlled.

The radio type of remote control for model aircraft typically uses the frequency of 27 MHz, and uses FM digital coding. Car locking/unlocking key remotes use a much higher frequency, typically 418 or 433.92 MHz, and this is subject to interference by defense department equipment and some radio amateur equipment. Modern cars use digital code systems that are much less likely to suffer from interference, but it is always reassuring if you can lock or unlock the car by the old-fashioned method of putting the key in the lock. Car keys also include a transponder chip that is passive (needs no battery) and which responds to a signal from the car when the key is put into the ignition switch. Early types used a transponder with a fixed code, but the modern version uses a code that is changed on each use (by a signal from the car) to make copying extremely difficult. All of this electronic gadgetry makes replacing a car key a very expensive problem.

Garage door controls have in the past used several frequencies, including the 28–30 MHz that often conflicted with model aircraft frequencies. The trend in Europe has been to standardize the frequency of 868.35 MHz, but many controls still in use work with other frequencies such as 818 or 433 MHz. The 418 MHz frequency is no longer used, but there could still be some around. All of this means that if you need to renew a garage door controller you will need to check with the original manufacturer.

Security Equipment

Definition

Under this heading we can group the devices that use **passive infrared** (**PIR**) detectors, with or without lights. The principle of PIR detection is that all humans and animals have a temperature higher than their surroundings (certainly in the UK), and this makes them net emitters of IR radiation that, though invisible, obeys the same physical laws as visible light. Though we cannot detect infrared with our eyes, there are several semiconductor devices that can and which can be formed into thin films. A typical material is gallium nitride, but there are many others.

A normal glass lens is not efficient for focusing IR, but the type known as a Fresnel lens (after its inventor) can be used. A Fresnel lens consists of concentric rings cut into a material that is transparent to the radiation (so that Fresnel lenses for light, as used in lighthouses, can be made from glass or plastics). The output from the sensor can be amplified in a conventional IC and used to operate lights or a burglar alarm. The sensitivity is of the order of 1°C above the surrounding (ambient) temperature, and the object needs to be moving at a speed of at least 10 cm/s. Maximum sensitivity is achieved if the moving warm object is traveling laterally, not approaching or receding directly.

PIR detectors are used in burglar alarm applications to reinforce the traditional magnetic sensors fitted to doors and windows. The usual format of a domestic burglar alarm permits two modes, with a partial mode that uses only the door and window sensors so that humans and animals can move around the house without setting off the alarm. The other mode is full, with the PIR detectors activated, and used when the house is empty.

Robotics

Definition

A robot is an intelligent mechanism, capable of making decisions from sensed data and acting on these decisions.

The word 'robot' has been misused. It is often applied to devices that are under radio control by a human, so that they are simply human-activated. Going one step on, the word is often used to describe the automatic vacuum cleaner or grass cutter that can find its way around obstacles. This is closer to robotic activity, but it requires a human to determine whether the carpet or the grass needs attention. A true robot device for these actions would determine for itself when the action needed to be carried out, and would recharge its batteries from a socket when it finished its action.

We are getting close to creating a true robot, but we're not quite there yet. The basis of all robotics work is the use of sensors to gather data, the use of microprocessors to interpret the data, and the use of transducers to carry out the actions. Most of the robot machines we know about (because there may be many that are military secrets) are industrial robots, working at repetitive tasks that cause humans to experience boredom that leads to carelessness. Another rapidly growing class is the domestic robot, such as the vacuum cleaner and the grass cutter.

Machines that we could class as near-robots are also used where great precision is needed (in surgery, for example) or where a human would be in great danger (defusing bombs or exploring hazardous areas). They are not totally genuine robots because a real robot would carry out the diagnosis for surgery or find the dangerous situation for itself. In other words, a true robot could replace a human, even if only in a limited area of activity. There is slow but steady progress being made to this objective, and some day we may find that Isaac Azimov's three laws of robotics (from as long ago as 1940) need to be applied. These laws were formulated as:

1. A robot may not injure a human being or, through inaction, allow a human being to come to harm.

2. A robot must obey orders given it by human beings except where such orders would conflict with the First Law.

3. A robot must protect its own existence as long as such protection does not conflict with the First or Second Law.

This is a long way from *Electronics Simplified*, but it is a well-mapped path.

Further Reading

Books

If you need more information on electronics, for design purposes, for constructional or for hobby interests, some of the books listed below may be of interest. In addition to these books, an immense amount of practical help is available from the catalogs of Futurlec in the USA or Maplin Electronics in the UK. The websites are www.futurlec.com and www.maplin.co.uk, respectively.

For detailed information on new devices, use Google or other search engines on the Internet.

Some very useful reference books in electronics are either out of print or difficult to obtain. They have been included in the list below because they can often be found in libraries or second-hand shops.

Note that some books that are listed here as published by Butterworth-Heinemann may have been reprinted or revised under the Newnes imprint.

Circuitry

Analog Circuit Design (Williams) Butterworth-Heinemann, 1998
Analogue Electronics (Hickman) Butterworth-Heinemann, 1999
Digital Logic Design, 3rd Ed. (Holdsworth) Butterworth-Heinemann, 2002
Bebop to the Boolean Boogie (Maxfield) Newnes, 2003
Digital Logic Gates and Flip-Flops (Sinclair) PC Publishing, 1989
Electronic Circuits Student Handbook (Tooley) Newnes, 1995
Electronics: Circuits and Systems (Bishop) Newnes, 2010
Higher Electronics (James) Newnes, 1999
Introduction to Digital Systems (Crisp) Newnes, 2000
Newnes Electronic Circuits Pocket Book, Vol. I, Linear IC (Marston) Butterworth-Heinemann, 1999
Newnes Electronic Circuits Pocket Book, Vol. II, Passive and Discrete Circuits (Marston) Butterworth-Heinemann, 1993
Practical Analogue Electronics for Technicians (Kimber) Newnes, 1997
Practical Digital Electronics for Technicians, 2nd Ed. (Kimber) Newnes, 1998
Practical Electronics Handbook, 6th Ed. (Sinclair & Dunton) Newnes, 2007

The Art of Linear Electronics (Linsley Hood) Butterworth-Heinemann, 1998
Troubleshooting Analog Circuits (Pease) Butterworth-Heinemann, 1993

Components

FPGAs: Instant Access (Maxfield) Newnes, 2008
Manufacturers' Databooks by Texas, RCA, SGS-ATES, Motorola, National
Semiconductor, and Mullard contain detailed information on semiconductors,
with many applications circuits.
Operational Amplifiers, 2nd Ed. (Jiri Dostal) Butterworth-Heinemann, 1993
Operational Amplifiers, 4th Ed. (Clayton) Newnes, 2003
Passive Components, 2nd Ed. (Sinclair) Newnes, 2001
Sensors and Transducers, 3rd Ed. (Sinclair) Newnes, 2001

Computing

The *Made Simple* series (various authors) Butterworth-Heinemann

Construction

Electronics Assembly Handbook (Brindley) Butterworth-Heinemann, 1993
Management of Electronics Assembly (Oakes) Butterworth-Heinemann, 1992
Newnes Electronics Assembly Pocket Book (Brindley) Butterworth-Heinemann, 1992
Soldering in Electronics Assembly (Judd) Butterworth-Heinemann, 1999
Starting Electronics, 2nd Ed. (Brindley) Newnes, 1999

Electronics and Computing Topics

Build and Upgrade Your Own PC, 4th Ed. (Sinclair) Newnes, 2005
Newnes PC Troubleshooting Pocket Book (Tooley), 1993

Formulae and Tables

Electronic Engineers Reference Book, 6th Ed. (Mazda) Butterworth-Heinemann,
1989 [A well-established standard reference book in the UK.]
GE Transistor Manual, 7th Ed. (General Electric of USA) [Even the early editions
are extremely useful.]
Newnes Radio and Electronics Engineer's Pocket Book, 18th Ed. (Brindley) Newnes,
1989 [Another excellent source of information, easier to get in the UK.]
Radio Amateurs Handbook (ARRL) Latest edition is dated 2008 [A mine of information
of transmitting and receiving circuits. An excellent British counterpart is available, but

the US publication contains more varied circuits, because US amateurs are less restricted in their operations.]

Radio Designers' Handbook, 4th Ed. (Langford-Smith) Iliffe, 1967 [A wealth of data, though often on vacuum tube circuits. Despite the age of the book, now out of print, it is still the most useful source book for audio work. A recent reprint (1997) from Newnes is still available.]

Reference Data for Radio Engineers (ITT). Howard Sams & Co. [Probably the most comprehensive data book ever issued, but now out of print. Can be downloaded in PDF format from www.pmillett.com/tubebooks/Books/FTR_ref_data.pdf]

Microprocessors

32-bit Microprocessors, 2nd Ed. (Mitchell) Butterworth-Heinemann, 1993 [A valuable guide to the multitude of 32-bit processors, including RISC types.]

For information to design and construction of modern microprocessors, the Internet is the most useful source of information, because books go out of date too rapidly in this fast-changing area of study.

Servicing

Electronics & Electrical Servicing, 2nd Ed. (Sinclair and Dunton) Newnes, 2007 [Two books for Level 3 and Level 3 of C&G 6958 Course in Electrical and Electronics Servicing.]

Test Equipment

Modern Electronic Test Equipment, 2nd Ed. (Brindley) Butterworth-Heinemann, 1990
Newnes Electronics Toolkit (Phillips) Butterworth-Heinemann, 1997
Oscilloscopes, 5th Ed. (Hickman) Newnes, 2004

Useful Websites

www.electro-tech-online.com
www.epemag3.com
www.allaboutcircuits.com

Magazines

The magazines *Elektor* and *Everyday Practical Electronics* (EPE) are intended for the constructor and provide a wealth of practical information.

Color Codes for Resistors

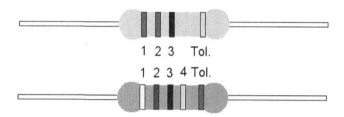

Figure	Color	Figure	Color	Figure	Color	Figure	Color	Figure	Color
0	Black	1	Brown	2	Red	3	Orange	4	Yellow
5	Green	6	Blue	7	violet	8	Grey	9	White

A similar scheme is used for capacitors with values in picofarads, but the scheme is often extended to indicate other features such as working voltage.

For the physical appearance of electronic components, see the websites devoted to electronics component photographs, such as:

> http://personal.ee.surrey.ac.uk/Personal/H.M/UGLabs/images/resistor_packages.jpg
> http://www.lessloss.com/images/ezpages/capacitors.jpg
> http://www.jestineyong.com/wp-content/uploads/2008/03/capacitor4.jpg
> http://www.garhttp://www.garwin-electronic.in.th/images/inductive.JPGwin-electronic.in.th/images/inductive.JPG

ASCII Codes

These codes for alphabetical and numerical characters are almost universally used in computing and other digital applications. The core of code illustrated here covers the numbers 32–127, using a maximum of 7 binary bits. This permits the use of parity for error checking within a single byte (8 bits). Extended ASCII uses numbers 32–255, permitting a wide range of characters such as accented characters, Greek, Hebrew, and Arabic characters, and other symbols, but these are not so strongly standardized. The best-known 8-bit set is the PC-8 set used on the IBM and clone computers.

The numbers have been given in both denary and binary forms. The table shows the UK ASCII set in which the £ sign is represented by 35. In the US set this number is used for the hash sign (#). The arrangement of numbers emphasizes the relationships that are built into the table, such as the single-digit difference between a lower-case letter (such as **a**) and its upper-case equivalent (**A**).

Denary	Binary	Character	Denary	Binary	Character	Denary	Binary	Character
32	00100000	(space)	64	01000000	@	96	01100000	'
33	00100001	!	65	01000001	A	97	01100001	a
34	00100010	"	66	01000010	B	98	01100010	b
35	00100011	£	67	01000011	C	99	01100011	c
36	00100100	$	68	01000100	D	100	01100100	d
37	00100101	%	69	01000101	E	101	01100101	e
38	00100110	&	70	01000110	F	102	01100110	f
39	00100111	'	71	01000111	G	103	01100111	g
40	00101000	(	72	01001000	H	104	01101000	h
41	00101001	)	73	01001001	I	105	01101001	i
42	00101010	*	74	01001010	J	106	01101010	j
43	00101011	+	75	01001011	K	107	01101011	k
44	00101100	,	76	01001100	L	108	01101100	l
45	00101101	-	77	01001101	M	109	01101101	m
46	00101110	.	78	01001110	N	110	01101110	n
47	00101111	/	79	01001111	O	111	01101111	o
48	00110000	0	80	01010000	P	112	01110000	p
49	00110001	1	81	01010001	Q	113	01110001	q
50	00110010	2	82	01010010	R	114	01110010	r
51	00110011	3	83	01010011	S	115	01110011	s
52	00110100	4	84	01010100	T	116	01110100	t

(Continued)

Denary	Binary	Character	Denary	Binary	Character	Denary	Binary	Character
53	00110101	5	85	01010101	U	117	01110101	u
54	00110110	6	86	01010110	V	118	01110110	v
55	00110111	7	87	01010111	W	119	01110111	w
56	00111000	8	88	01011000	X	120	01111000	x
57	00111001	9	89	01011001	Y	121	01111001	y
58	00111010	:	90	01011010	Z	122	01111010	z
59	00111011	;	91	01011011	[	123	01111011	{
60	00111100	<	92	01011100	\	124	01111100	|
61	00111101	=	93	01011101	]	125	01111101	}
62	00111110	>	94	01011110	^	126	01111110	~
63	00111111	?	95	01011111	_	127	01111111	(delete)

Index

Prisms, color television, 156–65
Probes, definition, 316–17
Production methods, CDs, 233–4
Program counter (PC), 244, 253–7
Program registers, definition, 243
Programming signals, 59, 205–8, 255
  see also Microprocessors
Programs, 243–61, 265, 274–9, 286, 293–5
  see also Drivers; Software
  definition, 243, 265
  drivers, 274–9, 293–4
  installation issues, 293–5
  languages, 286
  running, 253–5
Projection displays, 161
PSK *see* Phase shift keying
PSUs *see* Power supply units
Pulse repetition rate (PRF), 312–13
Pulses, 18, 20–2, 44, 53, 59, 86, 88, 90–2, 142–4, 149–54, 167–83, 199–201, 205–21, 244, 246, 257–8, 312–13, 315–16
  see also Clock…; Digital circuits
  actions, 21–2, 205–21
  definition, 18, 20, 22
  electronic equipment, 20–1
  leading edges, 20–3, 205–8
  timing, 20–2, 88, 90–2, 167–83, 199–201, 205–8
  trailing edges, 205–8
  triggering edges, 205–8, 315–16
  uses, 20, 53, 59, 149, 205–8
Pylons, 38
Python, 286

QAM *see* Quadrature amplitude modulation
QPSK *see* Quadrature phase-shift keying
Quadrature amplitude modulation (QAM), 302
Quadrature phase-shift keying (QPSK), 303–4
Quantization, 173–6, 240–2, 300, 318
  see also Conversion

definition, 175
Quartz crystals, 41–2, 209–10, 323–4

Race hazards, 205–6
Radar, 17, 108, 111, 139, 146, 148, 173, 199, 311–14
  block diagrams, 312–13
  definition, 311–12
  Doppler radar, 314
Radio, 10, 16–19, 29, 41–2, 51–2, 63, 69, 70, 77, 86, 89–90, 93, 95–117, 219, 307–9
  see also Digital…; Stereo…
  BBC, 103–4
  block diagrams, 105–12
  circuits, 63
  distortion figures, 63
  early radio receivers, 93, 99–100, 105–8
  first entertainment stations, 102
  frequency amplifiers, 70
  historical background, 95–114
  homodyne receivers, 111
  Internet radio, 116–17
  ionosphere, 96, 109
  Morse code, 100–1, 137
  supersonic heterodyne receivers (superhet), 108–12, 212–13
  tuning, 29, 41–2, 51–2, 89–90
Radio carrier waves, 10, 101–12
  see also Oscillators
Radio Corporation of America (RCA), 156
Radio transmitters, 16, 72–4, 77, 93, 97–117, 141–65, 171–3, 218–21, 307–9
  see also Vacuum tubes
Radio waves, 13, 16–19, 43–6, 72–4, 95–117, 307–9, 324–5
  see also Antennas
  definition, 16–17, 95–9
  hertz, 16–17
  modulation methods, 72, 100–17
  Oliver Heaviside, 96
  reflected radio waves, 108–12, 301–5
  speed, 95–9

uses, 16–17, 18, 95–9
Random-access memory (RAM), 217, 247–9, 250–3, 260–1, 263–4, 273, 287–8
  see also Dynamic…; Static…
  capacities, 249–52
  definition, 217, 247–9, 264
  ROM, 252–3
  volatility, 247–8, 249
Raster
  see also Cathode ray tubes
  definition, 139
RCA *see* Radio Corporation of America
Reactance, 35–7
Read-only memory (ROM), 217, 247–8, 252–3, 260–1, 263–5, 282–5
  calculators, 260–1
  definition, 217, 247–8, 264
  EEPROM, 217
  EPROM, 217
  RAM, 252–3
Read/write lines, control bus, 251–2, 263–4
Reading list, 179, 189, 329–31
Real-time issues, analog systems, 171–2, 227–8
Real-time operating systems (RTOS), definition, 285
Reception
  radio waves, 93, 97–117, 141–65, 307–9
  video signals, 141–65, 302–9
Rechargeable batteries, 7–8, 321–2
  see also Lithium-ion…; Nickel–metal hydride…
Recording heads, 125–6
Recording methods, 119–36, 169–71, 175–83, 223–42, 306–7, 318, 319
  see also Cassette; Cinema sound; Compact discs; Digital…; Disc…; DVDs; Noise; Tape…; Video…
  Blu-ray recordings, 131, 171, 224, 236, 306, 319
  DAT, 136, 170, 306
  future prospects, 136, 306–7

Printed in the United States
By Bookmasters